Go-Go Tools

Five Essential Activities for Leading Small Groups

Shigehiro Nakamura

Productivity Press • Portland, Oregon

This book is derived from *Small Groups and the Go-Go Tools*
by Shigehiro Nakamura and Hideyuki Takahashi, originally
published in Japanese by Japan Management Association, 1994

English language translation by J.H. Loftus for Productivity Europe, 1996

Revised edition ©1997 by QCD Innovation Technology & R&D Ltd., Tokyo, Japan
Revised U.S. edition ©1998 by Productivity Press, a division of Productivity, Inc.

All rights reserved. No part of this book may be reproduced or utilized in any form or by any means, electronic or mechanical, including photocopying, recording, or by any information storage and retrieval system, without permission in writing from the publisher.

Additional copies of this book are available from the publisher. Discounts are available for multiple copies through the Sales Department (800-394-6868). Address all other inquiries to:

Productivity Press
P.O. Box 13390
Portland, OR 97213-0390
United States of America
Telephone: 503-235-0600
Telefax: 503-235-0909
E-mail: service@ppress.com

Cover design by Mark Weinstein
Page composition by Lorraine Millard
Printed and bound by Malloy Lithographing in the United States of America

Library of Congress Cataloging-in-Publication Data

Nakamura, Shigehiro.
Go-go tools : five essential activities for leading small groups / Shigehiro Nakamura.
p. cm.
"Originally published in Japanese by Japan Management Association 1994"—Cip's pub. info.
ISBN 1-56327-200-8 (pbk.)
1. Small groups. 2. Teams in the workplace. 3. Decision-making, Group. 4. Management. I. Title.
HM1333-N33 1998
302.3'4—dc21 98-17033
CIP

£21-00

Coventry University

03 02 01 00 99 98 10 9 8 7 6 5 4 3 2 1

Go-Go Tools

LANCHESTER LIBRARY, Coventry University
Much Park Street, Coventry CV1 2HF Telephone 024 7688 8292

10 OCT 2000 CANCELLED
12 OCT 2005 CANCELLED
14 DEC 2000 CANCELLED
14 FEB 2002 CANCELLED
19 FEB 2002 CANCELLED
29 NOV 2002 CANCELLED
13 MAR 2003 CANCELLED
20 MAR 2003 CANCELLED

...ok is due to be returned not later than the date and
...d above. Fines are charged on overdue books

Contents

Publisher's Message

When small groups were introduced over twenty years ago in Japan, their main purpose was to heighten employees' sense of belonging to their companies and to create a culture of participation and involvement. Now their purpose has expanded to promote individual growth, create dynamic workplaces, and attain organizational objectives. By having managers entrust the solution of problems to those people who are most familiar with the work, companies are provided with solutions that are both practical and appropriate. Also, companies resolve problems more quickly and without conflict, while employees experience a greater sense of satisfaction and purpose in their work. Creative problem-solving activities of small-groups also help to develop the individual abilities of employees, which in turn strengthens the company. As Shigehiro Nakamura, the author of this book, says, "Small-group activities are an extremely important element of human resource development. I would even go so far as to say that developing people is one of their chief purposes."

Go-Go Tools—Five Essential Activities for Leading Small Groups is culled from the work of Shigehiro Nakamura, a Japanese consultant who for the past several years has been bringing some of the latest Japanese management techniques to Productivity Europe. Mr. Nakamura was encouraged to publish an English version of some of the more useful techniques, and the result is this book, which focuses on the use of go-go-tools for small-group activities in the manufacturing industry. (Go means five in Japanese—in this book, there are *five* tools and each has *five* steps.)

In *Go-Go Tools*, Shigehiro Nakamura provides five tools and procedures to help small groups use their own knowledge and ingenuity to autonomously solve problems and find solutions that are aligned with organizational objectives. The tools are; the pocket matrix technique, the volume-weight (VW) technique, the 5-why technique , the fact-finding (FF) technique, and the idea mapping technique. These tools will help small groups solve problems expertly and quickly in a rapid, continuous improvement cycle. The tools are practical because they are arranged in accordance with a given procedure. Each tool corresponds to a definite step of the problem-solving process—from identifying the problem to effecting the improvement. Anyone can use them because their purpose and methods are easy to understand and apply. Throughout the book Mr. Nakamura also provides important requirements, guidelines, tips, and procedures, such as TP management (total productivity management) to assist companies in developing, managing, and evaluating small-group activities so that they remain effective and aligned with organizational goals. He also stresses the role managers have to play if small-group activities are to succeed.

Chapter 1 deals with some of the problems companies have encountered in using small-group activities, such as incorrectly applying improvement techniques and forgetting that the aim of small-group activities is to obtain maximum result for minimum outlay. Mr. Nakamura discusses seven problems that can affect any company's small-group activities. A few of these problems are: laying too much stress on presentations, leaving the running of the program to the organizing staff while top management stands aloof, failing to calculate the activities' return on investment, long-term participation by inappropriate people, and having managers with insufficient practical experience organize the program. He also discusses the pitfalls of running a small-group program isolated from the day-to-day management of the company and shows several "incorrect" small-group styles that a surprising number of firms continue to use.

Chapter 2 discusses how companies should manage small groups and assess their results. Mr. Nakamura provides seven hints for solving small-group management problems. He discusses how managers should spell out the "What?" and the "Why?" but leave the "How?" up to the workplace groups; how to give training in the workplace and keep it practical, teaching each group only the tools it needs; how to skillfully connect shop-floor problems to management issues; how to make it clear that small-group activities are an important part of people's regular duties; how to skillfully draw out workers' knowledge of the problems; and why it is important to have management draw up standards at the job site. The chapter finally discusses key points of successfully running small groups.

Chapter 3 introductes the specific techniques of the go-go tools for use in management and small-group activities. Here you will learn the purpose of the five tools as well as how to apply the systematic, five-step problem-solving procedure. (All the tools are illustrated except for the FF technique.)

Chapter 4 presents two real-life examples of the application of the go-go tools—the use of the tools by truck drivers at a distribution center and a particular method of implementing improvements using video cameras. In the second example the author also discusses triple-speed kaizen—how to solve problems and effect improvements on the spot at three times the usual speed by condensing three steps into one simple step.

Chapter 5 describes a simple *5-sheet presentation* method to enable groups to communicate their achievements effectively to one another, which will help to raise their general level of technical ability. The essence of the method is to use a simple format and no more than five or so overhead projector transparencies to present information. The chapter also provides basic steps for compiling information for presentations; hints for drafting a good theme, or story, to hold the presentation together and keep it interesting; a simple technique for creating an effective presentation; a procedure for addressing a topic; three key points for speaking techniques; and finally, ways to make your presentation dynamic and interesting.

Chapter 6 covers some recent developments in TP management and its relationship to small-group activities. TP management helps you obtain a clear idea of the strategies you must implement—strategies that you develop by taking your overall objectives, breaking them down into individual objectives, checking them in terms of customer satisfaction, and incorporating the ideas and opinions of the company's employees. This chapter uses the *TP deployment diagram* to provide a simple explanation of the principles of TP Management. The TP deployment diagram is a visual application of the plan-do-check management cycle. It works by structuring the connections between macro-objectives and micro-objectives, which enables you to calculate the cumulative total of the micro-results achieved and compare this total with the macro-objectives. Managers will also learn how to boost the performance of small-group activities using a five-step evaluation—planning, trust, example, selection, and creativity—and learn six further requirements for good management. The chapter also provides managers with a five-step system for setting challenging targets and raising the level of their small-group activities.

Chapter 7 reviews some of the psychological and scientific principles of small-group behavior. The author discusses Maslow's five-stage hierarchy of human needs in the context of linking individual needs to social needs. He stresses the importance of creating a "dream" as the driving force in developing people. He discusses the Hawthorne Experiments of the 1920s that proved the existence of human motivation. He shares techniques that are used to improve team sports and tells how they can be used to strengthen and boost teamwork within the company. Through the McGregor Theory X (compel people to work) and Theory Y (respect people's needs for self-esteem), Mr. Nakamura shows why praising your workers is better than criticizing them. You will also learn about the 2-6-2 principle that explains the typical response of a group when something new has to be done.

Chapter 8 begins by giving a brief history of the zero defects movement and small-group activities in Japan. It then discusses how the changing business environment will impact small-group activities in the coming century. To guide the future development of small groups, a four stage concept is presented by which their progress can be charted and evaluated. The four stages are voluntary participation, self-actualization, contribution to performance, and participation in management.

We are grateful to Malcolm Jones, editor and publisher for Productivity Europe Ltd., for bringing the Japanese manuscript to bound book in the English language, to John Loftus for the translation, and to John Hardaker for the original illustrations. We especially wish to thank all those who participated in bringing this new version to bound book; Gary Peurasaari, development editor, for arranging and editing the book in its present format; Susan Swanson, production editor; Mary Junewick, copyeditor and proofreader; Lorraine Millard for typesetting and revised illustrations; Smith & Fredrick Graphics for new illustrations, and Mark Weinstein, for cover design.

Steven Ott
President & Publisher

Introduction

Over two decades have passed since small-group activities started up in Japan. During this time, the Japan Management Association (JMA) has done its best to keep the small-group movement active and to get small groups established, mainly through its zero-defect (ZD) campaign. These activities have now taken firm root in almost every Japanese company. People have however, become too used to them, and they are consequently beginning to deteriorate.

Approximately three years ago, a large number of industrialists took part in discussions of the JMA's proposed model for small-group reform, designed to breathe new life into the small-group movement. The aim of these discussions was to find the best way of promoting the kinds of small-group activities that will be closely tied to corporate objectives and that will directly affect corporate business performance. Various companies have been investigating this topic since that time, and the JMA has received an extremely large number of comments on the subject. Two of the more interesting observations are as follows:

1. If small-group activities are carried out in purely bottom-up style (that is, on the sole initiative of the people involved in them without being directed at attaining top-down objectives), the effort put into them will not be fully exploited and they will have very little effect on corporate business performance.
2. Although the so-called seven QC tools are used extensively by small groups, their application does not always lead to real improvement. People seem to end up simply playing with the techniques for its own sake. Often, a great deal of time is spent on using the tools to create attractive presentations after an improvement project has already been completed.

This book is based on the results of a large-scale survey of companies performed by the JMA on the future of small-group activities. This was done in conjunction with investigations carried out by the JMA's small-group activity committees into the current problems with small groups and what to do about them.

The book's principal aims are:

1. To suggest ways of promoting small-group activities that dovetail with corporate objectives in order to improve corporate business performance.
2. To indicate how to develop the individual abilities of a company's employees through creative activities and thereby strengthen the company as a whole.

Small-group activities are an extremely important element of human resource development. I would even go so far as to say that developing people is one of their chief purposes. They are a means of stretching employees' abilities at the same time as boosting the company's business achievements.

This is why this book focuses on human resource development. As Maslow's five-stage hierarchy of human needs indicates, effective management requires the establishment of a corporate "dream" that will accord with the personal dreams of individual employees. This forms the basis for the setting of ideal goals that the entire workforce will strive to attain. The aim is to have the whole company working as a team and experiencing the satisfaction of achieving its goals, while ensuring that everyone's efforts lead directly to the fulfillment of the company's overall business objectives. This is done by setting worthwhile improvement projects, during the course of which people develop their abilities, begin to see the underlying problems, and understand what they must tackle next. This leads to the setting of further objectives, which leads directly to the achievement of the company's overall business goals.

The JMA first proposed its model for small-group reform, designed to indicate the direction that small groups should take in the future, at a national small-group conference held in June 1992 to mark the 25th anniversary of the ZD campaign. This model defined small-group activities as activities designed to help an organization achieve its goals—goals carried out autonomously and continuously by small groups that try to solve problems and attain targets by pooling all their members' knowledge and ingenuity (see Figure I-1).

The JMA also specified the three key purposes of groups as follows:

1. *Promotion of individual growth:* to encourage personal growth through self-development and mutual enlightenment by respecting people's autonomy and providing them with opportunities to exercise their individuality and creativity.
2. *Creation of dynamic workplaces:* to build lively and energetic workplaces by mobilizing group synergy.
3. *Attainment of organizational objectives:* to contribute to the achievement of organizational goals by improving the quality of work through the reformation and refinement of working practices.

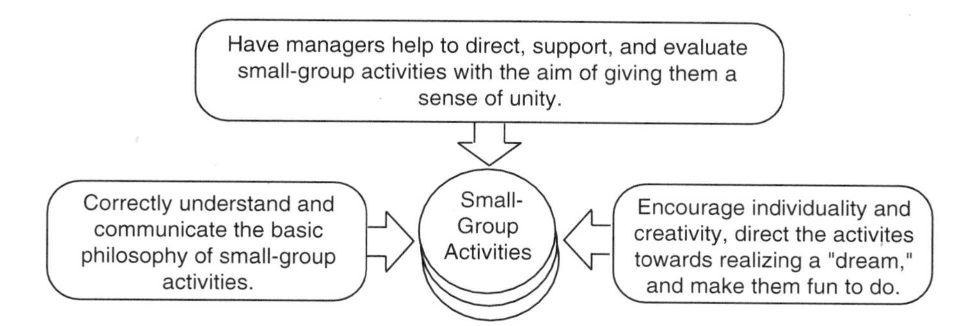

Figure I-1. Guiding Principles to Pool the Knowledge and Ingenuity of All Members

CHAPTER 1

Problems with Conventional Small-Group Activities

As mentioned in the introduction, many companies are running into trouble with their programs of small-group activities. This chapter discusses the different types of problems encountered. Among other issues, we will be examining:

- Laying too much stress on presentations.
- Leaving the management of the program up to inexperienced staff.
- Incorrectly applying improvement techniques such as the seven QC tools.
- Forgetting that the aim of small-group activities is to obtain maximum result for minimum outlay.

The chapter includes a checklist for determining whether the small-group activities practiced at your company have become divorced from the business realities, and a description of four typical cases in which this has, in fact, occurred.

THE HISTORY AND THE MANAGEMENT BENEFITS OF SMALL GROUPS

When small groups were first introduced into Japanese industry over twenty years ago, their main purpose was to heighten employees' sense of belonging to their companies and to create a culture of participation and involvement. This was effectively accomplished. Later, the focus shifted towards the solution of problems experienced by employees in their workplaces, and it was at this point that the activities began to make a major contribution to the management of the companies that had introduced them. The main benefits they produced were as follows:

1. Since the improvements effected through small-group activities were introduced by the people most familiar with the work, they were usually both practical and appropriate to the situation.
2. The activities helped people experience a greater sense of satisfaction in their work, and they began to study improvement techniques assiduously. Ordinary workers became able to solve many of the types of problems that previously had been dealt with by trained engineers.
3. Managers entrusted the solution of increasing numbers of problems to the people in the workplace and, as a result, more and more of them were

solved without conflict. This meant that greater and greater amounts of authority could be delegated, and the ability of the workplace to manage itself increased.

Because of these benefits, and the fact that small-group activities increase people's sense of belonging and help them develop their individual skills and abilities, many senior executives hastened to introduce the activities into their companies. They were also probably motivated by seeing the very positive effects small-group activities were having on the performance of other companies that had introduced them.

As a result of this process, small-group activities have become well established at large numbers of Japanese firms. In many cases, however, it is blindly assumed that they will automatically result in bottom-line benefits, and no real attempt is made to tie them in to corporate business objectives. Consequently, they have become divorced from the real work of the company, and the people engaged in them are more concerned with the application of improvement techniques than with the results achieved. When this happens, it should be taken as a warning signal indicating that the activities have gone off course and need to be reviewed from the management standpoint.

THE RITUALIZATION OF SMALL-GROUP ACTIVITIES

This section describes seven problems that can affect any company's small-group activities. If five or more of them apply, it should be taken as a sign that the activities are becoming ritualized and are beginning to emphasize form over content.

The basic question we must ask ourselves is, "What and who are we pursuing our program of small-group activities for?" If the aims of the program are not clearly defined there will be a tendency to put the teaching of techniques before the achievement of results. When the program's organizing staff takes this approach, the members can easily become complacent, content merely to be running the activities even though those activities are not really very productive. This is a particularly prevalent problem, and it is essential to break out of this situation whenever it arises.

If we take the view that a company's small-group activities should have a direct and measurable effect on its bottom-line results, we can see the following seven types of problems with small-group activities as currently practiced.

Problem One: Presentations Are Full of Enthusiasm, but the Content of the Activities Is Remarkably Thin

Presentation techniques have been well honed and are extremely good. The presenters are very slick, but their real skill is only in creating a dramatic spectacle. In the worst examples, the activities only last for a week or a month, while the presentations go on for three months. In some cases, preparation for presentations can take a similar length of time, despite the low level of activities and the lack of any significant results. This is emphatically *not* how a program of small-group activities should be run.

This situation is often glaringly obvious when observing small-group presentations. For example, cause-and-effect diagrams may be presented to explain setup reductions, even though such improvements should really be described in the form of a time analysis. In this case, the diagrams are no more than a sort of optional extra bolted on as an afterthought. The problem here is that the presentation is the focus of attention, while the actual activities take a back seat. Presentations like these also foster the wrong attitude in the people attending them, as people become overimpressed with the presentation skills displayed and often even applaud the approach instead of criticizing it for its excessive emphasis on form over content.

Problem Two: Top and Middle Managers Leave the Running of the Program to the Organizing Staff and Appear Only at Presentations

Of course this may be to some extent inevitable, since it is the organizing staff who is responsible for running the program. However, if upper management omits to set goals, the organizing staff is unable to judge how well it is doing. Small-group activities are sometimes referred to as being autonomous or self-directed, but this is no excuse for senior management to adopt a laissez-faire attitude and neglect to define objectives or to suggest how the activities should be managed.

Nevertheless, even when they are mostly left to their own devices by upper management, the organizing staff often feels that it has been given a job to do and runs the program with tremendous zeal. Eager to do their best, staff members get closely involved with the small groups and listen enthusiastically to their problems and ideas. However, the more effort the organizing staff makes, the more the senior managers are left out and lose interest. In the end, everything is left up to the organizing staff.

Problem Three: The Focus Is on Presentations, and Small-Group Activities Are Started Only After the Problems Have Been Solved

Much analytical work using tools such as cause-and-effect diagrams and Pareto diagrams has absolutely no practical benefit, since it is done after the cause of a problem has already been found. When we take a close look, we see that analytical work is not, in fact, being used to track down causes. A cause-and-effect diagram, for example, may contain a wealth of information and impress people with how much a small group knows about a situation. However, if the information is not actually used for identifying the cause of the problem, the diagram is no more than a device for showing off the group's knowledge.

Most problems arising in the workplace have only a single cause and require only a single solution. All we need to do is find the one true cause of each problem and take action to eliminate it. In principle, cause-and-effect diagrams should be used to track down this one true cause. However, in reality, this does not often happen; such diagrams are frequently prepared as an afterthought for the purpose of presentations. A whole range of meaningless additions like this may be introduced. This sort of trend has been apparent of late in Japan.

In organizations that have studied this problem, small groups using tools such as cause-and-effect diagrams repeatedly ask themselves, "Why are we doing this?" This approach has led to the skillful application of the method, and good results are obtained by fully investigating the true problems while writing down relatively few causes.

Sometimes, small-group activities start only after a problem has already been solved. The group prepares arrays of spectacular graphs and tables because they know that the standard of their presentation is being judged, not the standard of their solution. In the worst cases, despite the meager content, the techniques that are used make attending the presentation feel just like a trip to the theatre!

Problem Four: Activities Are Run Without Giving Any Thought to Return on Investment

In my view, small-group activities should be assessed on the basis of their cost benefits. We should evaluate both training and the activities themselves in terms of whether or not their goals justify the amount of time and money invested in them. Classroom training is a particular case in point, since instructors are often satisfied simply with having taught the techniques, while the students sit there bemused, with no idea of how to actually use them. Losing sight of the purpose of the training, the instructor then starts teaching the students the mathematics behind the tools.

Many instructors tend to value the tools themselves more than their application. This makes the students even more confused about how to apply the tools in actual problem-solving situations, so they begin studying tool techniques in the classroom. This in turn involves them in deep discussions and in drawing up mountains of statistics and graphs. What is more, they are taught far more techniques than they can cope with.

I myself doubt whether the statistical techniques used by academics are actually very useful for making improvements through small-group activities. Essentially, if we have a problem, the question is how to find its one or two true causes. Then, all we need to do to solve the problem is to eliminate whatever we regard as those causes. When students are given all sorts of irrelevant information and have a great time simply studying the seven QC tools or any other set of improvement techniques, then the creation of data and charts becomes the aim and the investigation of causes is forgotten.

This is a fairly common situation, and many companies complain that, although their people train in improvement techniques, they never come to grips with any really worthwhile issues. Presentations often give the game away, revealing that the kind of training-focused approach described in the previous paragraphs is going on. In such a situation, presentations feature a whole range of techniques not actually used in practice. This is just a waste of time. It is essential to give the kind of training that will yield benefits commensurate with the time and money invested.

Problem Five: Fixed Times Are Allotted for Group Activities, but They Produce No Useful Outcome

Here, activities are organized, but they do not produce any noticeable results. The groups' direct supervisors generally take no interest at all in what is going on, and the upshot is that merely engaging in the activities at the appointed times becomes an end in itself. The thinking behind this uncontrolled state is something like, "If we keep on running the program, we should eventually see some results."

In this kind of situation, the company's managers usually do not know what they are trying to achieve through their small-group activities or where their problems lie. In many groups in companies like this, the principal focus ends up on issues such as how to get everyone talking, how to collect opinions, how to draw clever diagrams, or how to write impressive reports. In such a situation, small groups naturally remain ineffective despite regular times being set aside for their activities.

Nor does any improvement in business performance occur under these conditions. If managers are making comments like, "Results should come through at some time or other," or "We'll see some benefits once the group members develop their potential," it means that managers are not providing enough guidance. Groups often select topics their direct supervisors do not approve of, so the supervisors take little active interest and eventually end up saying things like, "We're not getting any results so we might as well give up." However, the existence of an established system for conducting the activities at designated times makes it difficult to give up, and the situation drags on indefinitely with nobody formally complaining about what is going on.

Problem Six: Too Much Importance Is Given to Full Participation, Even When the Issue Being Addressed Concerns Only a Handful of People

Really, the right way to solve a problem is to bring together just those people who possess the required knowledge and skills to solve it, and then go ahead and solve it as quickly as possible. However, in the name of full participation, the members of a problem-solving group are often chosen before the topic has been chosen, and even people who have nothing to do with the issue are included. These uninterested parties have to be brought into the discussion whether or not they have anything useful to contribute. The effective members then have to spend too much time listening to them and the activities have a hard time getting off the ground and producing any useful results.

Meanwhile, a group leader's main concern becomes to simply *involve* everyone, and the issues of solving the problem and effecting improvements, which are in fact the whole point of the activities, end up taking a back seat.

Problem Seven: People Without Practical Experience Run the Program and Lead the Groups

In the worst case, an inexperienced new recruit may be told to run the program on his (or her) own. Still relatively ignorant about the workplace, he is given these duties apparently for his own education. Conscious of his position, he sets about trying to achieve results by whatever means he can. However, given his lack of practical understanding, he inevitably tends to lean too much towards the theory of managing small groups rather than on practical application. As a result, the focus is on how to set up the system and how to make the groups work rather than on the results achieved.

Such a person is of course unable to give the small groups any specific guidance on what to do. The best he can do is to invite instructors in from the outside and to organize training sessions. However, the more this becomes his job, with the goal of training as many people as possible, the harder it becomes to address the essential questions. As a result, even the training ends up serving no useful purpose. It is important for groups to be led by people who are able to draw on their own practical experience. In my view, the most effective way of ensuring that small-group activities lead to significant bottom-line results is to operate a system that promotes effective leadership by appointing people who have successfully engaged in and directed small-group activities. Only they are competent to provide on-the-spot instruction and guidance relating to specific problems.

Summarizing the above seven problems, we can say that small-group activities take off in the wrong direction because of:

1. Overemphasizing presentation techniques at the expense of results.
2. Leaving the running of the program to the organizing staff, while top management stands aloof.
3. Adding on the preparation of presentation materials as an afterthought to the activities themselves.
4. Failing to calculate the activities' return on investment.
5. Ritualization of the activities, with no attention paid to them by direct supervisors.
6. Long-term participation by inappropriate people.
7. Organization of the program by managers with insufficient practical experience.

These problems result from losing sight of what and who small-group activities are run for. If any of these points apply to your program, then a review is urgently needed.

RUNNING A SMALL-GROUP PROGRAM ISOLATED FROM THE DAY-TO-DAY MANAGEMENT OF THE COMPANY

I believe there are also some companies that have introduced small-group activities for reasons of external image. There must also be firms that have delegated an organizing staff to begin group activities from the viewpoint that, "It's produced results in other companies, so why not try it here too?"

Running a small-group program isolated from the day-to-day management of the company leads imperceptibly to an uncontrolled situation in which the activities become totally unrelated to the company's business results. Many companies are now facing this problem. The managers of such firms frequently say things like, "The trouble is, small-group activities do nothing to address the most important management issues. Something must be done." They then copy what other companies are doing and devote all their efforts to teaching problem-solving techniques without first working out which problems need to be solved. A surprising number of firms continue in this way with stereotyped training in the vain hope that it will eventually lead to some concrete results.

The JMA survey mentioned in the Introduction shows that this situation is, in fact, a common one. Table 1-1 gives some typical examples. The four companies in the table introduced small-group activities because other firms were running them, or because they just appeared to the companies to be a good idea.

At the K company, large-scale presentation meetings are frequently held, and an excellent small-group training manual has been produced. The organizing

Table 1-1. Examples of Small-Group "Diseases"

Company and Small-Group Style	Principal Problems
K Company (a building society) People at this company are practicing small-group activities without really knowing why.	1. The organizing staff members have no real answer when asked why they are running the small-group program. Their only response is, "Everyone else is doing them, so we brought them in here too." 2. Small-group activity is dominated completely by presentations. 3. The activities are conducted at set times with instruction manuals handed out. These manuals are excellent. A lot of money is spent, but the employees do not do anything useful and the results are minimal. The sole concern is to motivate people to take part in the activities.
N Company (steel manufacturer) Small-group activities have been running for a long time here, with the focus exclusively on quality improvement. P Company (pipe maker)	1. Production conditions are so different from what they were before that the number of topics needing to be addressed has gradually declined and the point of the activities has been lost. 2. Presentation meetings have continued for a long time, but they are more for appearance than for any real value. Top management is uninterested and does not even attend. When a presentation on a particular workplace has finished, everyone connected with that workplace leaves the room without staying to hear the remaining presentations. 3. The organizing staff changes every 1-2 years. New staff members are impatient for their next posting and do not put much enthusiasm into the activities.
This firm is excellent at producing posters that analyze the current situation and show the progress of the activities. E Company (parts assembler)	1. Twenty or more groups have been active for three years, but only four projects have been completed. 2. Progress charts are plastered all over the walls of the company's large canteen. 3. The progress of the small-group activities is always discussed at monthly section managers' meetings, but the managers only receive reports and take no further action. A large part of these reports consists of supervisors' comments.
Impressive posters are stuck all over the factory walls, and virtually the whole range of improvement tools is used.	1. Training funds are allocated on an annual basis, and the budget is huge. 2. The groups themselves produce hundreds of colorful, time-consuming posters. 3. Despite all the posters, the failure rate has stayed at 2 percent for the past two years. The failure reduction rate is in the order of ppm.

staff members are proud of it and even boast that they could probably sell it to other companies. However, an inquiry about the content and aims of the small-group program produces no clear explanation. Furthermore, results are not very evident. Already several years have gone by with the company putting great efforts into the program in the belief that they are bound to get some results if they keep at it long enough.

The N company has been pursuing various activities for some time now, but again to little effect. The problem arises from the fact that the activities have been run by a mix of different people who have not been very enthusiastic about the job. At this company, presentations are the only part of the program done properly. Middle managers attend the presentations but apparently do not get involved in leading the day-to-day activities. Top managers are completely indifferent. As might be expected, the program has consistently failed to produce any significant results.

The P company has been running a program of small-group activities for three years and has over 20 groups, even though it has only 150 employees. The company claims that all these groups are active. Judging from the wealth of information on group progress posted in the canteen and the carefully prepared monthly reports on group activities, its claim does at first sight appear justified. All this effort is most impressive. However, a closer examination reveals that, although there are plenty of supervisors' comments, the data consists mainly of charts and tables showing analyses of the existing situation and plans for future action.

In fact, only four of the twenty or more projects undertaken over the three years of the program have actually been completed. This represents an extremely poor return on investment.

In the E company, great importance is attached to training, and the organizing staff believes in spending large amounts of money on it. As a result, huge sums are spent on educating small groups. Neglecting their ordinary work, the group members put great energy into drawing up charts and tables. The workplace walls are virtually invisible under all the posters that have been stuck up. The JMA's reaction to this is to ask whether the company is a paper mill or poster producer rather than a product manufacturer! Nevertheless, despite all the impressive activity, the failure rate has remained unchanged for the past two years, at about 2 percent. A lot of effort is being made but no real improvements are being achieved.

If situations like these continue—that is, small-group activities hung up on appearance, presentation and form—anarchy will spread, and the activities will eventually become totally disconnected from the company's business performance.

As mentioned earlier, these problems were identified from the responses to the JMA survey as well as from comments made by individual companies approaching the JMA for advice. The JMA wondered whether any companies had succeeded in solving these problems. They felt it would be wrong to simply publish a list of bad examples, so they resolved to review what successful companies were doing with their small-group activities and to try to point others in a similar direction. As a first step, they decided to get companies to check whether they were suffering from the small-group "diseases" described above, and to take remedial action if they were. The next chapter describes the cures they effected.

Chapter 2

Managing Small Groups and Assessing Their Results

The most important point in small-group activities is to ensure that the people actually doing the work will be constantly on the lookout for problems and will keep on improving the quality of their own work—by themselves. This raises the professionalism of the employees in their work, as well as having a desirable effect on the management concerns of quality, cost, and delivery. It is therefore essential for management to look carefully at the topics selected by the group and to turn the spotlight on small-group activities to find out how to guide the groups in the right direction. With this in mind, this chapter gives seven hints for dealing with some common problems.

SEVEN HINTS FOR SOLVING SMALL-GROUP MANAGEMENT PROBLEMS

Hint One: Managers Should Spell Out the "What?" and the "Why?" but Leave the "How?" Up to the Workplace

Managers should set the topics to be addressed, and the small groups should look at these and select those within their capabilities. Alternatively, they may select a topic that is outside their range of competence and take it on while receiving the necessary coaching. In this approach, the people in the workplace can exercise their ingenuity in finding solutions (the "How?") to problems (the "What?") set in accordance with top-down policy.

This approach is needed in order to break out of the situation in which management concerns itself too closely with the "How?" and neglects to tell people the "What?" and the "Why?" Although the small groups may become conversant with some highly effective problem-solving methods, the activities will lose their sense of direction because, as mentioned before, people become so wrapped up in the methods used that they forget all about the activities' purpose.

Hint Two: Give Training in the Workplace, and Keep It Practical

The best way of leading small groups is to appoint people with practical experience or people who have already had some success with small-group activities as instructors. Instruction that is delivered in idealistic or general terms or that does not relate to the problems the group is tackling will always be a waste of time. Moreover, problems and their causes occur in the workplace, so it is essential to address the issues and do any necessary training right where the problems are actually happening. If the training focuses on real issues, with the aim of immediately utilizing everything that is learned, then the techniques will be quickly understood and results will soon appear.

Hint Three: Teach Each Group Only the Tools It Needs

Once people have a clear idea of the problem, the only thing left to do is to work out how to solve it. I believe employees are perfectly capable of solving most problems for themselves, without using any particularly sophisticated techniques, provided that they know exactly what the problems are. According to one well-known management consultant, line graphs and bar charts are all that is needed as a basis for workplace management. This is taking things to extremes, but provided that the important problems are prioritized and the groups know which ones they should be wrestling with, they will go ahead and find their own ways of solving them. All that is required is to drop a hint or two at the right time. It is unnecessary and even counter-productive to teach each and every technique; it is more important to first explain exactly what the problem is, and then to teach the required tools as and when they are needed. I believe that this is what training is all about.

If I may give a rather far-fetched example, it is not much good telling people to practice driving Formula One racing-cars if you want them to climb a mountain. On a less exaggerated scale, you need to equip people with different gear for hill-walking than for rock-climbing.

People may have a hard time trying to apply a tool you have given them but, if it is the right one, they will gain valuable insights into the problem-solving process through doing so. If you want a problem to be solved quickly, it is essential to define the objectives clearly and supply the appropriate tools at the appropriate times.

Hint Four: Skillfully Connect Shop-Floor Problems to Management Issues

Many important management issues cannot be resolved by front-line employees on their own. For example, increasing a company's turnover would require that a whole host of factors be addressed. Such an objective could only be achieved by splitting it up into separate elements such as improving telephone skills, improving fax communication, improving customer service, and so on. Similarly, if people on the shop floor are simply ordered to increase their productivity, they will not really know how to. We need to give them specific tasks such as correcting sporadic machine stoppages or eliminating defects by installing simple measuring devices. When we do this, improvements take place, the work flows more smoothly, and productivity naturally increases.

This is why I consider it so important for managers to start by asking the people in the workplace what their problems are, then to classify these in terms of their effect on quality, cost, and delivery performance, and then to discuss the reasons for various possible courses of action. Having done this, they should then set the goals to be pursued so that concrete action can be taken and the problems can ultimately be solved.

Hint Five: Make It Clear That Small-Group Activities Are an Important Part of People's Regular Duties

Effective leadership by direct supervisors is the key to the success of small-group activities. At one time, various activities were pursued under the title "self-directed activities."

Certainly, one of the benefits of small-group activities is the way in which they improve communication, promote teamwork, and develop better interpersonal relations. However, while these things are undoubtedly important, they are not the basic objective. Moreover, in themselves they are not something to spend too much time on. The greatest amount of collective satisfaction is generated when a group suddenly realizes, "Yes! We can do this!" through the process of solving a problem, which in turn improves relationships among the group members. The well-known maxim "success breeds success" is very true. When a group succeeds in effecting an improvement early on, the satisfaction its members feel makes them want to solve the next problem. This is the real significance of small-group activities.

This is why we should think of small groups almost as problem-solving project teams, and should assign them topics through the organization's formal management structure to be part of their regular duties, rather than letting them select their own. Once small-group activities have been positioned as part of employees' regular work, the activities will of course need to be supervised. Group members' direct supervisors must develop their groups' abilities by meeting with them regularly in the workplace to ask how things are going, whether good progress is being made, and whether they have everything they need. Supervisors can also provide appropriate suggestions whenever teams get stuck. This is the only way to ensure that small-group activities run well.

Hint Six: Management Must Skillfully Draw Out Workers' Knowledge of the Problems

Shop-floor workers usually know what the production problems are, but may not have enough technical knowledge to describe them or set about solving them. At one plant, for example, products were being damaged during the manufacturing process. The job involved grinding a material on a whetstone, which was used with a grinding liquid. Technical staff went to the workshop and drew up a cause-and-effect diagram in an attempt to identify the possible causes of the damage. Their investigation was based on the belief that either the machine was acting up or there was something wrong with the whetstone, and they were determined to find out what was happening. Something was certainly scratching the work but, despite their search, they could not find the cause anywhere. Just as they were about to give up, one of the workers piped up, "This is the reason." He dipped his hand into the grinding liquid and mixed it up a little, whereupon scratches immediately appeared on the work. He had known the cause of the problem all along!

In fact, the man had previously made the following suggestion: "Because particles in the grinding liquid scratch the product, we should be particularly careful about recovering them. We should install baffle plates in the feed tanks and make sure that we clean the liquid thoroughly before it's recirculated." However, his boss at that time had rejected this proposal out of hand, saying, "There's no point in that." After receiving this rebuff, the worker had given up, feeling that he had better keep his mouth shut.

The workers on the spot often know the cause of a problem but their supervisors sometimes interpret their comments as complaints and reject what they are saying. As long as they continue to do this, problems will remain unresolved.

As this example shows, it is essential for managers to skillfully elicit workers' knowledge of what is wrong, monitor the problems, relate them to quality, cost, and delivery performance issues as described earlier, and proceed to solve them one by one. Although the results may initially appear insignificant, if such efforts are built up over a period of time they are certain to eventually bring about marked improvements.

Hint Seven: Draw Up Standards at the Job Site

The basis of manufacturing is, of course, standardized working practices. People in offices often write up standards and then give instructions for the employees in the workplace to follow them, but such standards frequently end up being deliberately posted where nobody can see them. In other cases, the standards are so hard to follow that a completely different job method is actually used. This sort of thing is a good topic for small groups. If standard procedures are followed, the work can be done better, faster, and more reliably. The most important thing is to identify the best ways of working and ensure that they become firmly established.

Once the workplace has mastered the process of setting down the best working practices in the form of standards and then continually improving the standards, all kinds of activities will go more smoothly, basic conditions will be maintained, and bottom-line results will start to come through.

ACHIEVING GOOD RESULTS WITH SMALL-GROUP ACTIVITIES

A great deal can be learned from companies that are achieving good results with their small-group activities. An important point about these companies is that its managers are usually highly skilled at leading their entire organization in the direction they want it to go. By "management," I mean not only the company's managing director and site managers, but also its middle managers and the direct supervisors of front-line employees.

To achieve good results with small-group activities, it is essential to set up a system whereby, under the guidance of these managers, improvements appear naturally. Results will be achieved across the board if small groups are assigned definite topics but are left free to decide how they will tackle them. All companies that have posted excellent results through their programs of small-group activities are convinced that, without these activities, they could not have made the structural improvements they have, in fact, achieved.

I would now like to pick out the key points from examples of successful companies that are cleverly balancing the process-oriented approach with the achievement of results, and that are adroitly combining the top-down and bottom-up styles of management. Since this book is about small groups and the go-go tools and "go" means "five" in Japanese, I have selected five key points for successfully running small groups.

Key Point One: Promote Group Activities As Projects

Rather than adopting the problem-solving approach (people in a workplace looking for what is wrong with the existing situation and solving the problems thus identified), small-group activities should be run in the form of projects addressing definite topics. Under this system, management sets a topic and selects the most suitable people for solving it. Then, after the topic has been set, the group directs its own activities, aiming to complete them as quickly as possible. In today's conditions, when workplace staff are frequently rotated and products are continually being modified, training and project activities need to be closely interlinked and must be accomplished expertly and quickly.

The need for speedy results means that functional boundaries must be crossed and people unconnected to a particular workplace must be brought in to work on its projects. Long-term issues are carefully broken down into individual tasks, which are then addressed in turn. Furthermore, all the people gathered must discuss the topic as a united, democratic, problem-solving team, regardless of rank.

One company that has succeeded with this approach calls its activities "three-day focused improvements." Another successful company sets the topics and then appeals for people who want to be involved. A third excellent company brings together all the interested parties from various workplaces.

All three of these companies are using this project-based approach to generate stunning results. What is more, it is the main driving-force behind their improved factory profitability, higher quality, and shorter lead times. This is because management initiates the projects, setting topics that mainly concern critical processes and technologies. Also, since there is no permanent project organization, and project teams are formed and disbanded as and when they are needed, there is no need to set up any special positions or pay structures for team members. When forming teams, the only question managers have to ask is, "What sort of group do we need to solve the problem?" The result is activities that generate quick and effective solutions.

Key Point Two: Develop an Effective System for an "Idea Bank"

It is all very well to say that managers must set topics for small-groups to tackle, but what if they have no ideas? To avoid this situation, it is important to gather ideas from a wide variety of sources. For example, in regard to personnel matters, we could collect comments such as, "Our new employees are not learning fast enough." In regard to working practices, we could collect questions such as, "What level of standardization have we reached?" or "If we display the procedure like this, even people taking over from us will be able to produce high-quality work," or "What if we did the measurements this way?" Suggestions such as, "How about promoting a little more cooperation and exchange of knowledge between production workers and technical staff at the design stage, and making this particular component this way?" also provide important hints for coming to grips with problems.

A matrix like that shown in Figure 2-1 is drawn up and used as an always-open storage bank for comments, questions, and suggestions like those just described. This "idea bank" system is an excellent technique for amassing large amounts of information on problems, difficulties, or things that managers want done in a certain way. The idea bank can act as a rich source of ideas for topics that need to be addressed.

As Figure 2-2 illustrates, to move from ideas to their practical application, part of the idea bank's contents are drawn off in accordance with company policy, and selected ideas that mesh with this policy are placed in a smaller "storage tank." These are then taken by small groups as their topics, and they work through these topics one after another.

Whenever people take some action, improvement ideas are bound to occur to them. If everyone in the organization is in the habit of accumulating his or her

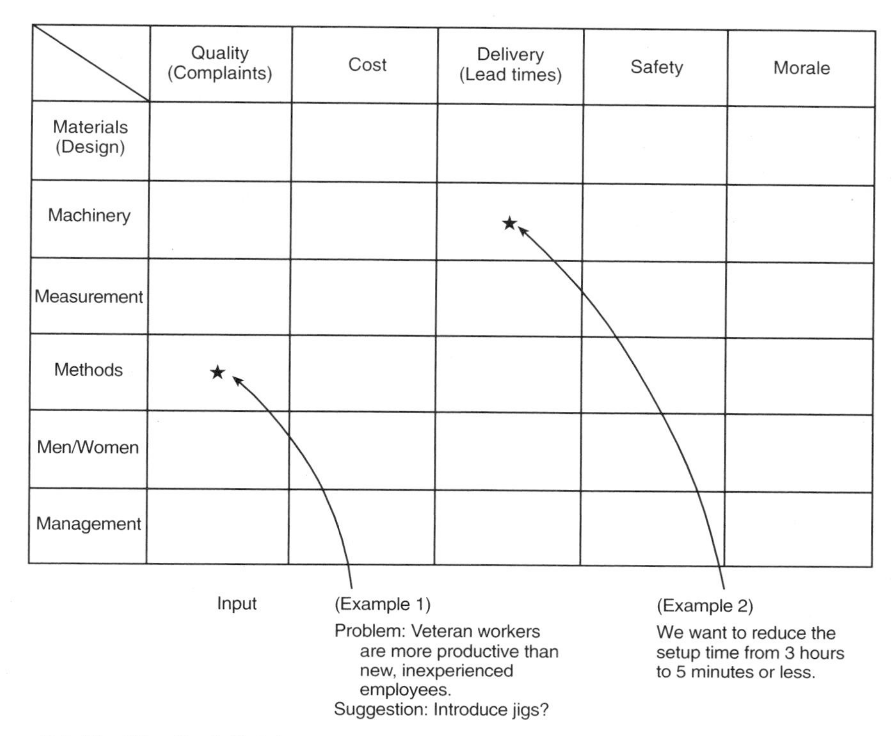

Figure 2-1. The Idea Bank Matrix

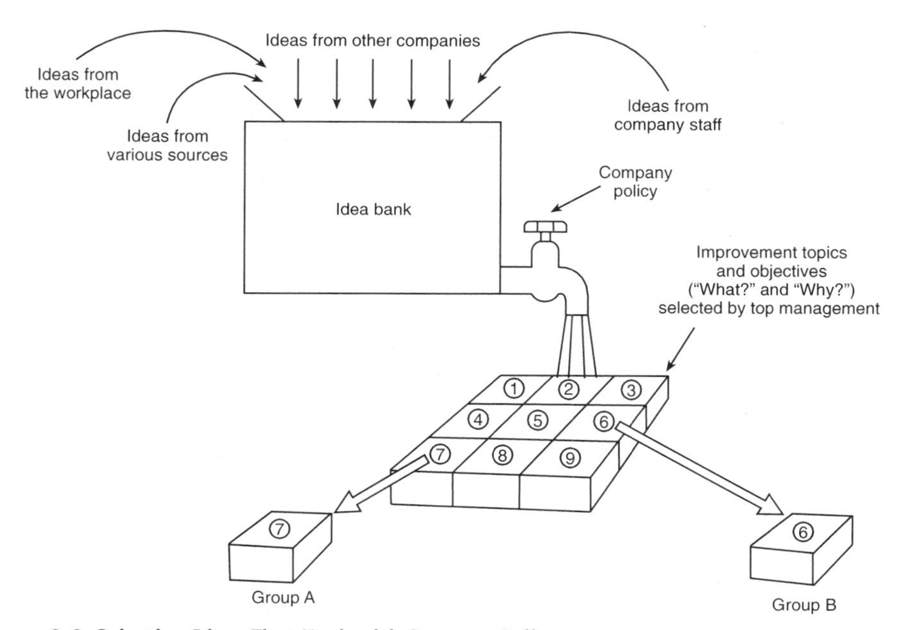

Figure 2-2. Selecting Ideas That Mesh with Company Policy

thoughts in the way described, the storage tank will soon be full of valuable ideas for use in future improvement projects. The system will prove even more effective if the idea bank is laid open for all to see. A more detailed description of the technique will be given later.

Key Point Three: Make Full Use of "Visual Management"

The JMA advocates the management system known as total productivity management (TP management), widely regarded as the most transparent management system presently in existence. A more detailed description of TP management is given in Chapter 6, but it can be thought of in simple terms as a system whereby the entire management of an organization is represented by a matrix. The matrix is constructed with improvement policies and guidelines up the side and Q, C, and D management objectives across the top. The contents of an idea bank like that just described are examined and stored in the appropriate cells of the matrix. The most effective ideas are then extracted from this pool, with a bias towards the company's current objectives. As Figure 2-3 shows, the management system and the details of the actions being undertaken can be seen at a glance.

This procedure makes it possible to show systematically where the management focus lies by providing a holistic view of all the company's policies, process problems, QCDSM (quality, cost, delivery, safety, morale) relationships, and so on. This in turn makes it obvious why each particular topic must be addressed and which problem in which process it relates to. In other words, it enables the effect of individual strategies on the achievement of the company's overall objectives to be assessed in an extremely visual way.

The system illustrated in Figure 2-3 is designed to clarify the relationships and priorities among company issues, objectives, and policies. If a program of small-group activities is backed up by a system along these lines, and is tailored to the company's particular circumstances, then groups will be formed on the basis of a shared understanding of their roles and responsibilities and the importance of the tasks they are undertaking, and problems will be more readily solved. A system such as TP management is indispensable for running a business effectively in today's unforgiving business climate.

Key Point 4: Look for New Ways of Approaching Topics

Every technique has its limitations, so it is important to progressively change one's improvement approach. When kicking off an improvement campaign, the first step is usually to get everyone to agree on the need for it. We start by tackling the people (men/women) issue of motivation. If this goes well, we then begin to think about how to achieve what we want—in other words, we move on to consider our methods. Everyone unites in activities such as tackling quality

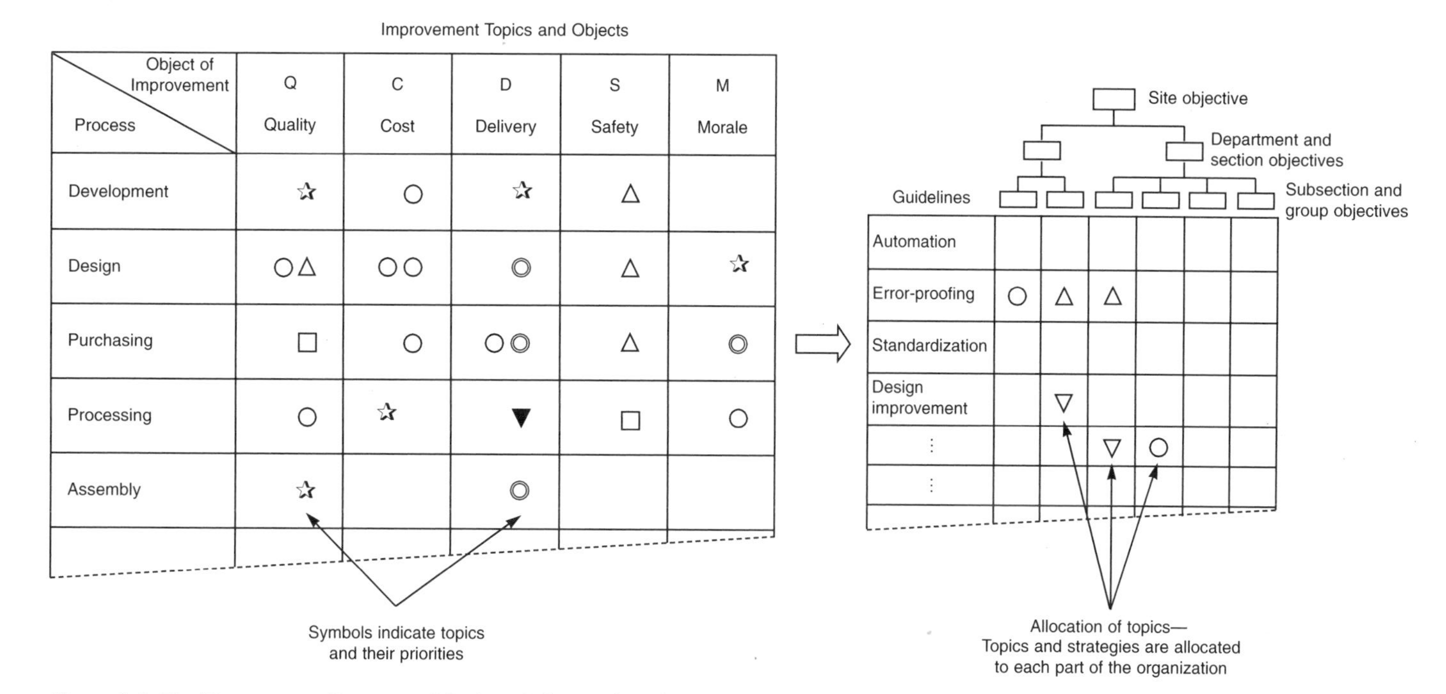

Figure 2-3. The Management System and Actions Being Undertaken

problems or making better products by improving the techniques used to make them. This is the second stage, in which we attempt to create faultless products through the use of QC techniques. In the third stage, we try to devise more efficient ways of doing the work.

To achieve this, we proceed to eliminate waste by employing industrial engineering techniques such as time analysis. Eventually, we find ourselves producing excellent-quality products with less labor than ever before. In other words, we have succeeded in raising the efficiency of the operation without sacrificing the quality of our products. If we then review the tasks carried out by the workers, we will probably find that they have become far simpler than they were.

At this stage, somebody will probably come up with the idea of measuring whatever is measurable and introducing jigs or other mechanical devices to further simplify the work. In other words, we raise the level of sophistication of the work a step higher. In short, making the work as simple as possible constitutes the groundwork for introducing mechanical aids.

As more and more mechanical aids are introduced, we start to think about mechanizing the work. For example, we might decide to introduce computers, fax machines, and word processors to mechanize administrative tasks. As the level of technology is gradually raised in this way, various types of low-cost automation devices and industrial robots begin to be introduced, and the process equipment itself may even be remodeled. This is the stage at which the fruits of people's knowledge and expertise are consolidated in the form of hardware.

Yet there are limits to this, also. When a certain stage is reached, it becomes impossible to push costs down any further. When people begin to have thoughts such as, "If the product shape were modified a little, it could be made more quickly," or "With the product like it is at the moment, this machine requires a great deal of labor, and the process is also too time-consuming," it will not be long before they start wanting to direct their improvement efforts to points even earlier in the production sequence to try to improve product designs. This is generally called the "design-in" approach, or design for manufacturability. In this approach, a far-sighted view is taken, and production or office workers are included in design projects right from the outset with the aim of making improvements at a point as far upstream in the production flow as possible. Although this can be very effective, the higher up the production chain improvements are directed, the longer they take to implement and the greater the expertise they require.

I call this process of simultaneously improving and upgrading by progressively shifting the focus of improvement and gradually raising the level of technology the "5-step principle of QCD improvement." Figure 2-4 illustrates this principle. It is important for managers to realize early on that there is a limit to

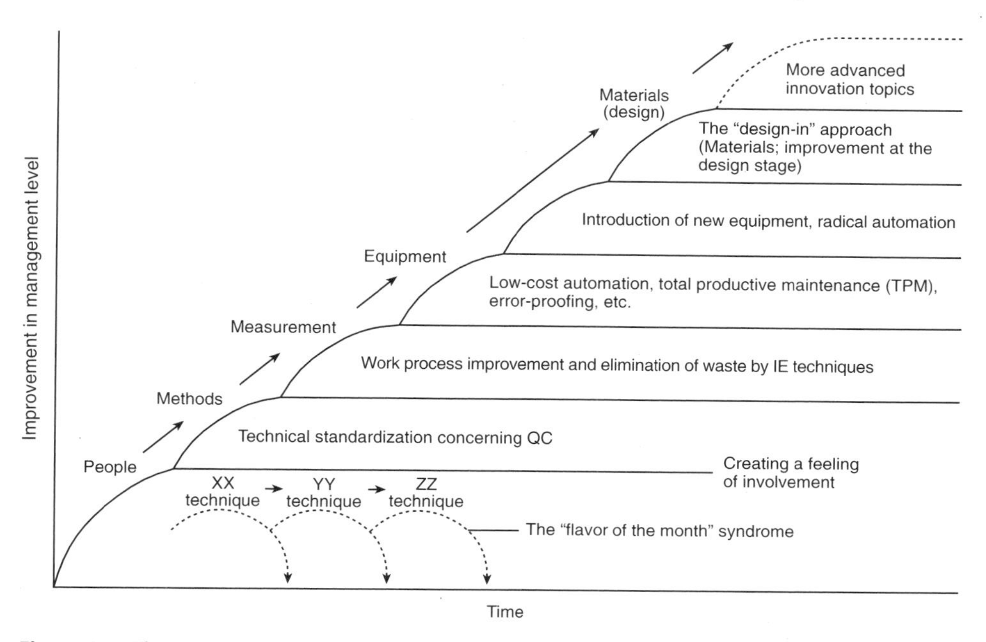

Figure 2-4. The 5-Step Principle of QCD Improvement

the effectiveness of any management technique or improvement approach. As soon as they see that a particular small-group activity has outlived its usefulness, they should skillfully point the group in a different direction.

Key Point Five: Always Explain the Purpose of Improvements

Small-group activity topics should, in my opinion, always be linked either to customer satisfaction (CS) or employee satisfaction (ES). While internally-focused improvements may bring rewards in terms of better financial results for the company, they often do not benefit the customer in any way, and the company does not see any increase in its orders.

This is why it is so important to look at our companies' activities from the outside and ask ourselves whether our customers are really delighted by the improvements we have made. This is the essence of CS-focused small-group activity.

At the same time, it is important to set topics that relate to ES. In one example, the people installing a computer system at a distribution company were trying to prevent form-filling mistakes. However, despite a thorough investigation, no good proposals were found. They then decided to try dealing with the topic by adding an ES-focused approach. The outcome was a system for reliable form-filling by a simple double-checking procedure.

Introducing the ES factor in this way makes it possible to view things from a fresh angle and a slightly broader perspective. In many cases, improvement initiatives fail because staff give high-handed instructions without consulting the employees, or simply provide a list of faults for correction.

Recently, however, there has been a significant move towards introducing the ES-oriented approach, including assessing processes relating to one's own. This means not just limiting topics to solving problems in one's own workplace and eliminating faults relating to those problems, but also re-evaluating the work from someone else's viewpoint, so that different ideas emerge. The ES approach is an important slant for managers and supervisors to introduce when leading small groups.

In this chapter, I have reflected on the faults of small-group activities identified in the previous sections, and have described some specific ways of obtaining good results. I strongly urge readers to follow the above suggestions if they want to achieve a close linkage between their companies' small-group activities and business performance while ensuring that the activities undertaken make good sense and meet with unanimous approval.

CHAPTER 3

The Go-Go Tools for Revolutionizing Small-Group Management

This chapter describes some specific techniques for use in management improvement and small-group activities. Small groups do not actually need a very large number of tools in order to effect practical improvements in the workplace. In fact, I believe that the five tools described in this chapter will be ample for almost any contingency. The set of methods described here comprises a systematic, step-by-step problem-solving procedure newly developed and applied by the JMA.

The JMA has entitled this group of techniques the "go-go tools." This name was chosen because the word "go" means "five" in Japanese and there are five tools with five steps in each tool. The name go-go also conveys a sense of urgency, energy, and excitement (as in go-go dancing!). The purpose of the tools is to allow small groups to solve problems expertly and quickly in a rapid, continuous improvement cycle by applying each tool in accordance with the given procedure. The tools have been arranged so that each corresponds to a definite step of the problem-solving process—from identifying the problem to effecting the improvement—and they are, consequently, extremely practical. I believe that anyone is capable of using them because their purpose is clear and they are simple to understand and apply.

The following is a brief introduction to the five tools:

Go-go tool no. 1: The pocket matrix technique

The purpose of this tool is to identify current problems and create a bank of ideas. It is a method of systematically gathering together and organizing as many potential improvement topics and ideas as possible (see Figure 3-1).

Go-go tool no. 2: The volume-weight (VW) technique

This is a technique for facilitating the selection of topics to be tackled by assessing the seriousness of problems and ranking them in relative importance. It is a tool for performing a primary screening of the issues with the aim of concentrating resources where they are most needed and achieving the maximum result for the minimum effort. Small groups use this tool to select the topics they will actually tackle (see Figure 3-2).

Go-go tool no. 3: The 5-why technique

This is a problem-analysis technique designed to pin down the true causes of problems. Once the problems have been identified, the next step is to track down their causes. The aim of this tool is to find the one true cause of each problem (see Figure 3-3).

Go-go tool no. 4: The fact-finding (FF) technique

Although the cause of a problem may have been found, the problem must still be quantified. This technique is used to display and analyze the facts about problems in an easily understood form right where the problems are happening (there is no figure for this tool).

Go-go tool no. 5: The idea mapping technique

This is a technique for selecting the best strategy for solving a problem once it has been identified and measured. It is a method of comparing and assessing a stock of problem-solving ideas from various angles, color-coding them, displaying them in easily-understood form, and selecting the best proposal (see Figure 3-4).

GO-GO TOOL NO. 1: THE POCKET MATRIX TECHNIQUE

The pocket matrix technique is a method of collecting and organizing problems. Its purpose is to create a bank of ideas about people's worries and concerns, the improvements they would like to see, and so on. It is used at the preparatory stage before a small group actually settles on a specific topic to tackle.

Generally speaking, skillful problem-solving takes place in the following three steps:

1. Identify the problem.
2. Find a method of solving it.
3. Apply the method in order to solve the problem.

The first question is therefore, "What is the problem?" The pocket matrix technique outlined in Figure 3-1 on page 26 has been designed to help answer this question. The aim of the technique is to collect and classify the problems by organizing them into a matrix in terms of the production inputs (basically, the 5 Ms; that is, men/women, methods, measurement, machines, and materials) and the production outputs (QCDSM; that is, quality, cost, delivery, safety, and morale). The problems recorded in each cell of the matrix are then examined to see what improvements need to be made in that particular area.

Let us examine the axes of the matrix a little more closely. The production inputs (the 5 Ms plus an optional sixth M for management or an I for information) are plotted up the vertical axis, and people's improvement proposals are examined to determine which of these inputs they relate to. Since money is also an input, it too could be added to the matrix as an extra item if we are considering cost reduction. However, it is probably best to start with the basic 5 Ms + I, as shown in Figure 3-1.

Meanwhile, the production outputs or results are plotted along the horizontal axis. Here, we consider what effects will be produced by the 5 M + I measures we might want to implement. Quality (Q) can mean either quality improvement or the reduction of complaints and claims. A third item (rework) could also be added here. Expanding the matrix in this way can be an effective method of addressing issues such as building in quality via the process.

The next item on the horizontal axis is cost (C). Topics in this column would relate to matters such as cost reduction, productivity improvement, and so on. Cost items could include fixed costs such as labor costs and capital depreciation in addition to variable costs such as energy costs, materials costs, component costs, and so on. Specific cost categories such as these could also be entered into the matrix to assist us in identifying improvement topics.

The next item on the horizontal axis is delivery (D). This represents time and could include items such as work times, production times, finishing times, setup times, preparation times, and so on. Again, it helps us identify improvement proposals if specific categories such as these are included in the matrix.

The next item is safety (S). Here, we look at the matrix to see what safety improvements we could introduce. These would naturally include measures to reduce difficult, demanding, and dangerous work, and could also include environmental protection measures. Any ideas to do with items such as these would be entered in this column.

The final column is headed morale (M). This column is used for checking what should be done to boost employees' enthusiasm and make their jobs more pleasant. Finding ways of creating a more relaxed work environment and making people's jobs easier by improving their training is an effective method of boosting morale. Morale could also refer to external morale; in other words, the morale of the local community. Protecting the environment and serving society are of course important concerns for any company, and items relating to these could also be included in this column of the matrix.

The pocket matrix technique is an extremely useful improvement tool. It is very important to collect as many improvement suggestions as possible and to use this kind of matrix to record them systematically.

Production Outputs (QCDSM)

Production Inputs (5Ms + Information—I)	1 Quality	2 Cost	3 Time/ Delivery	4 Safety	5 Morale
1 Materials (+ Design)					
2 Machines					
3 Measure-ment					
4 Methods					
5 Men/ Women					
Information					

Figure 3-1. Go-Go Tool No. 1: Pocket Matrix Technique

How to Apply the Pocket Matrix Technique

Each of the five columns in Figure 3-1 is divided through the middle by a dashed line. Each cell of the matrix is thus divided into two halves; one for managers to record their ideas and one for nonmanagerial employees to record theirs. The best way of gathering the ideas is to have people write their suggestions on self-adhesive notes and then stick the notes onto the appropriate half of each cell of the matrix.

Step 1 for Tool No. 1

If we want to make an improvement, we first need to identify a problem or find an issue to address. No improvement activity can begin without a topic to tackle, so we must gather various ideas in order to decide on such a topic. For example, in applying the pocket matrix technique, people could write the following kinds of concerns on their cards:

"This work is tiring and difficult; it should be improved in the following way..."

"Inexperienced workers have a hard time learning the more difficult tasks."

"I have to watch what's going on all the time or nothing gets done."

"I can't even take my holidays without worrying about what's happening while I'm away."

"I can't relax because I keep thinking the machine will start producing defectives if I leave it for a moment."

"I could operate another machine as well as this one if I didn't feel that I have to keep constantly watching this one."

Alternatively, people could write down things they would like to see happen, for example:

"I'd like to try this myself."

"I'd like to work in a place where there are no breakdowns and no defectives."

"I'd like to reduce our setup times to the following figure..."

"The administrative work would flow much more smoothly if this communication link were set up."

"Things would definitely go according to plan if we considered this in advance."

In this way, the pocket matrix technique is used to collect notes of all the things that people actually doing the work are bothered by now, or would like to see improved in the future.

When applying this tool to either office work or to the production floor, we also gather people's opinions and thoughts along the lines of:

"We want to make such-and-such a job enjoyable for everyone."

"I want to become such-and-such a person."

"We want to reach such-and-such a level."

"I want to learn such-and-such a job quickly, so I can do it as well as my more experienced colleagues."

"Our boss says we ought to carry out such-and-such a procedure this way, but I think we ought to do it this way."

As before, statements like these are written on individual self-adhesive notes and are stuck onto the appropriate cells of the matrix. Ideas, problems, worries—any topic is acceptable. Such problems or worries can be turned into improvement themes by adding ideas as to how they might be resolved.

The first step in using the pocket matrix technique is thus to get managers and shop-floor workers to put their heads together to think about the kinds of issues that small groups ought to tackle. The use of the matrix facilitates this process.

Step 2 for Tool No. 1

The next step is to take the problems and ideas collected in each cell of the matrix in Step 1 and spread them out in the cell (if it is large enough) or on a separate sheet of paper, so that all are clearly visible. It does not matter what method is used as long as all the statements are visible to everyone. Everyone then pools their ideas, using the topics already written down as brainstorming ideas, and continues to fill the pockets with the ideas and opinions generated.

Step 3 for Tool No. 1

In the next step we investigate other workplaces to see whether they have any better ways of doing things or have identified some good improvement topics. We then examine these closely and decide whether we can apply them to our own workplace. Using them as hints, we add other usable topics to our growing collection of ideas. Finding as many potential improvement topics as possible by visiting other departments in our company is extremely helpful for the next step—that of selecting which proposals to implement. Because we are visiting other departments within our company, there should be no problem about secrecy. Other departments' good ideas should be available for use in our own department. We could also look at other factories belonging to our own company or even go to other companies' facilities in order to find the best ways of doing things, and then use these in our own workplaces. We could also use things that we have studied in seminars, and so forth, as a further source of ideas. In this way, we keep adding as many ideas as possible to the matrix.

Step 4 for Tool No. 1

In this step we decide which of the suggestions looks most promising and examine these one by one in order to estimate their benefits. Naturally, we want to select the topics that will give the maximum result for the minimum input of labor and other resources. The most effective approach is to examine the areas of the matrix that are of greatest priority for the management of the company, see what ideas have been collected in these areas, and select the topics to be implemented from among these.

Step 5 for Tool No. 1

The final step is to examine those topics that have been screened out and have had their benefits estimated in the previous step, and select the ones that we are actually going to implement. As already mentioned, the best topics for small groups to tackle are those that will give the maximum results for the minimum time and effort.

By following this procedure, it is generally possible to build up a stock of improvement topics that would take approximately two years to work through. The topics to be implemented in each period can then be selected from the matrix. Once the small groups start their activities, many more new topics will occur to them, and these should also be recorded on the matrix whenever they are thought of. The matrix thus acts as a continually updated bank of improvement ideas.

A healthy stock of improvement topics is a great asset, since it forms the basis for future improvement activities. The most fundamental thing in small-group management is to start with a list of specific topics. It is important to find the best ways of getting managers and front-line employees to work together to identify topics of mutual concern that will improve their company's bottom-line performance. Likewise, it is extremely important in running a program of small-group activities to start by collecting together every possible improvement idea, including things that people have noticed by themselves and things that they have learned about from seminars or from visits to other companies.

GO-GO TOOL NO. 2: THE VOLUME-WEIGHT (VW) TECHNIQUE

When embarking on any type of improvement activity, we must first find out exactly how serious the problems that we have identified are. The VW technique is a graphic method of displaying the relative importance of problems. As Figure 3-2 shows, it is a way of visually displaying the problems in order of their seriousness.

Step 1 for Tool No. 2

The first step in the VW technique is to quantify the extent of the problems. For example, there may be problems with quality, cost, delivery, or the time required

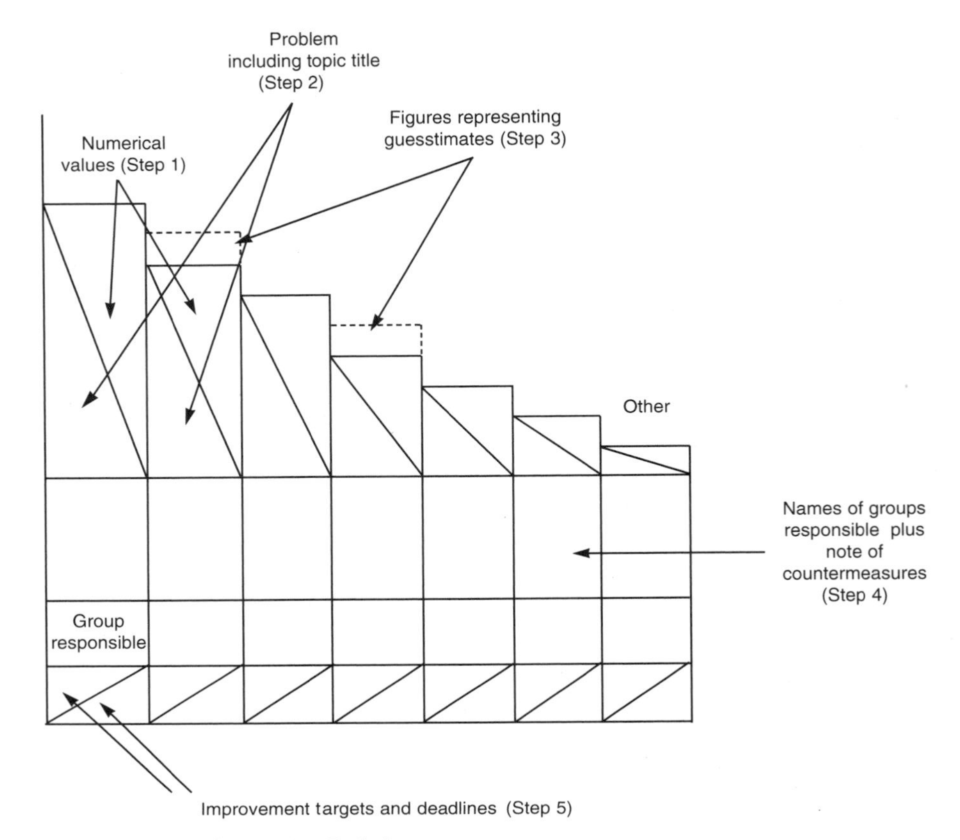

Figure 3-2. Go-Go Tool No. 2: VW Technique

to execute the work. We start by plotting on a block graph all those problems that can be quantified, starting with the most serious. We then divide each block into two parts by a diagonal line as shown, and record a numerical value indicating the seriousness of the problem in the upper half of the block. The graph now shows how serious the problems are.

Step 2 for Tool No. 2

Next, we write a description of the problem in the lower half of each block. This should also be done for unquantifiable problems. The graph now indicates what each problem or topic is about, as well as its seriousness in numerical terms. When describing the problem, it is important to state the specific trouble that has produced the numerical value indicated in the top half of the block. This should be done in the form of a complete sentence, not just with an isolated noun or verb.

Step 3 for Tool No. 2

In the next step, we further examine unquantifiable problems. Such problems might include complaints that the work is dangerous, dirty, or too physically demanding. With matters where life would become much easier or safer for the workers if the problems were resolved, or where there are other problems that are difficult to express in numerical terms, a figure should be estimated and placed within dashed lines. Sometimes, in the case of service industries, people say things like, "We could really please our customers if we did such-and-such." This kind of thing may not be capable of being expressed in numerical terms, but it can be added to the block graph in the form of a dashed line. Another kind of comment that can be included on the diagram in this way might be, "The problem could be prevented in advance if we were given the information more quickly." The VW technique allows the importance of even difficult-to-express topics to be demonstrated clearly and visibly.

Step 4 for Tool No. 2

Once we have displayed the problems on a diagram like this in visible form, and it is obvious which are the most serious ones, we think about moving ahead to the implementation stage. Starting with the most serious problem, we discuss who is going to tackle it. Having decided this, we make a note of which group is responsible in the space provided under the appropriate block. We then repeat this for each problem on the graph. The purpose of this step is to decide which group is responsible for which problem and show it on the graph.

Step 5 for Tool No. 2

The final step in the VW technique is to decide on a deadline for the completion of each topic. The purpose of this final step is to set a time frame in which problems should be solved and show it on the diagram.

We do not, at this stage, decide how we are going to solve the problems. Having used the pocket matrix technique to broadly classify the problems, the purpose of the VW technique is to determine how serious they are, rank them according to priority, give them a numerical weighting, select the topics to be tackled, and decide who is going to tackle them and by when they should be solved. In other words, the first two go-go tools are concerned with sorting out and prioritizing the problems, not with solving them. We do not enter the problem-solving phase until we reach tool no. 3.

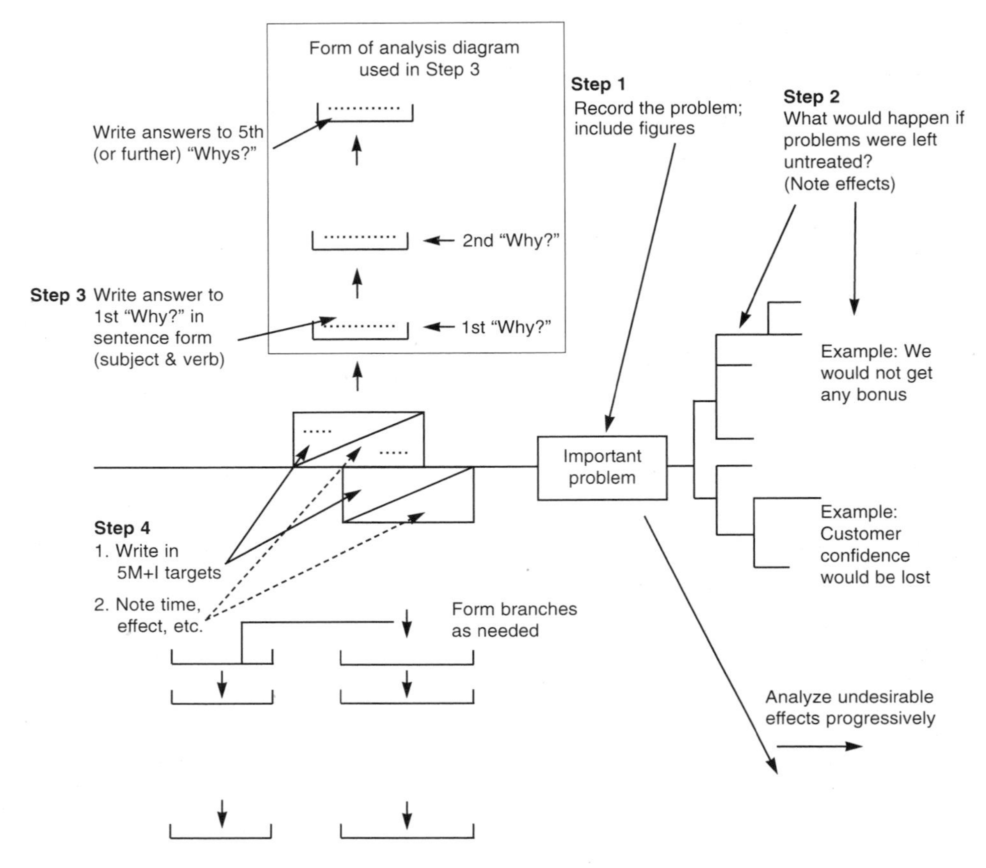

Figure 3-3. Go-Go Tool No. 3: 5-Why Technique

GO-GO TOOL NO. 3: THE 5-WHY TECHNIQUE

The purpose of this tool is to identify the true cause of a problem by repeatedly asking "Why?" (generally five times in succession). The 5-why technique is similar to the 5W1H technique (identifying the true causes of problems by asking the questions "What?" "Where?" "When?" "Why?" "Who?" and "How?") and the H-H technique ("How much?" and "How to?"), and is based on the shop-floor approach that employs the principle of the "three actuals" (go to the actual workplace, examine the actual objects in it, and look at what is actually happening there). The purpose of the 5-why technique is to get right to the heart of a problem and track down its real causes. Figure 3-3 shows specifically how the technique is used.

Step 1 for Tool No. 3

The first step is to record the problem in the box marked "important problem" in the center of the diagram.

Step 2 for Tool No. 3

The next step is to identify the seriousness and priority of the problem and its relationship to other factors. If people do not understand why they should address a particular issue, they will never really come to grips with it. A key purpose of this technique is to clearly demonstrate the seriousness and significance of the problem. We repeatedly ask the question, "What would happen if the problem were left untreated?" and write the answers on the right-hand side of the diagram.

Let me give an example of how this works in practice. We might ask the question, "What would happen if we did nothing about this kind of complaint?" The answer might be, "Our customers would get angry." Again we ask, "What would happen if we did nothing about it?" and this time we note, "Our customers would stop ordering from us," or "Our customers would lose confidence in us." Once more we ask, "What would happen if we did nothing about it?" and we realise, "Our company would not receive any orders." If we keep on asking the same question, we finally obtain the answer, "Our company's turnover would decrease." We thus use the right-hand side of the diagram to progressively analyze the undesirable effects of leaving the problem unsolved.

We could do a similar analysis by asking the question, "Why would it be wrong to leave the problem untreated?" We can use this procedure to consider the effects of leaving the problem unsolved on society, on our customers, and on ourselves. If people had previously been thinking that the problem was not really serious and did not need to be solved, this procedure helps to change their minds and makes them see that serious trouble could develop if nothing is done.

As a result of this step, the seriousness and significance of the problem are understood. This type of analysis gets people thinking that they really must do something to solve the problem. Once people's awareness of the problem has been heightened in this way, we shift our activities to the actual place where the problem is occurring, and try to identify its true cause by using the principle of the three actuals (go to the actual workplace, examine the actual objects in it, and look at what is actually happening there). From Step 3 onwards it is very important not to sit around in a meeting room but to get out into the workplace. Always go to the place where the problem is occurring and check the real facts.

Step 3 for Tool No. 3

In Step 3 we ask the question, "Why has the problem occurred?" When we have found the answer (that is, the primary cause of the problem) we write it on a card. This is the first step of the 5-why procedure. We then ask "Why?" again and find the secondary cause. We continue to ask "Why?" in the same way, finding one cause after another while checking the facts in the workplace. Generally speaking, we continue this process through five successive stages.

To make this process clearer, let me give an example. Imagine that a certain manufacturing process in which a workpiece is turned on a lathe is producing defective products. Our first question is, "Why are the defects occurring?" Our answer might be, "Because the lathe suddenly stops." Our second question is, "Why does the lathe suddenly stop?"

Our answer might be, "Because the fuse of the motor blows." Our third question is, "Why does the fuse blow?" If we stop the process of tracking down successive causes at this point, the only action we will take will be to change the fuse whenever it blows, and the problem is likely to keep occurring. To prevent this from happening, we might come up with the idea of replacing the fuse with one of a higher rating, but this would not be a good way of solving the problem.

Let us ask "Why?" once more—"Why does the fuse blow?" The answer might be, "Because it is subjected to an overload current." We then ask, "Why is it subjected to an overload current?" On investigating the workplace, we might come up with the answer, "Because the motor does not turn smoothly." We then ask, "Why doesn't the motor turn smoothly?" and we might guess, "Because it is not lubricated well enough."

Let us assume that we check this out and discover that we guessed right. On asking, "Why isn't it lubricated well enough?" we might reply, "Because not enough oil is reaching the motor." On asking, "Why isn't enough oil reaching the motor?" our attention might be drawn to the oil filter and we would probably then remove the cover of the oil filter and check it. We might then find the answer to our question: "Because the oil flow stops at the filter."

On asking, "Why does the oil flow stop at the filter?" the answer might be, "Because the filter is blocked." If the filter really is blocked, this is, of course, the cause of the problem and is the reason for the defects. We then look at the filter and ask, "Why is the filter blocked?" and we might conclude, "Because there is a lot of dirt and foreign matter around the filter."

Pressing on, we think about what a filter is for. An oil filter's essential function is to separate foreign matter from oil. Let us assume that there is so much foreign matter collected on and around the filter that the oil cannot get through it. We know right away that this is an undesirable situation, and we think very hard about what we must do in order to prevent this cause of defects from arising in the future. We realize that we have to devise some method of keeping contamination away from the filter.

One proposal for achieving this might be to remove the foreign matter by gravity settling before the oil reaches the filter. The oil passing through the filter would then be much cleaner, and foreign matter would be less likely to collect on and around the filter. This would probably be the least expensive and most reliable method of solving the problem. Not only would it prevent the fuse from blowing and obviate the need to replace it, it would also probably put to rest more expensive suggestions for solving the problem, such as replacing the motor with a more powerful one.

Not only has the cause of the defect been eliminated; because we have understood the root cause of the problem, but we have also been able to institute an inexpensive and effective countermeasure.

In this way, we use the 5-why technique to successively identify the facts about the problem in the workplace, repeatedly asking "Why?" in order to track down the real cause of the problem. Once this true cause has been discovered, we take action to eliminate it and thereby prevent the problem from occurring again.

As shown in Figure 3-3, the answers to the five "Whys?" are usually noted one by one on the diagram above the central horizontal line, but the procedure can be repeated below the line if required. The line of questioning may also branch off in different directions. Once the final cause has been identified, it is examined to see whether it is a human problem, an equipment problem, and so on (in other words, which of the 5 Ms + I it relates to). The results of this analysis are also noted on the diagram.

The cause-and-effect diagram is still an extremely popular method of tracking down the causes of problems. However, it is rather cumbersome to draw and, because it contains only isolated words, it is not a very effective way of approaching the true causes of problems. In the 5-Why technique, the answers to the successive "Whys?" are recorded in sentence form. This makes it far more likely that the true cause of the problem will be correctly expressed. Since, as noted above, the targets of countermeasures are also recorded on the diagram in terms of the 5 Ms + I, the diagram also facilitates the implementation of countermeasures. The true cause of the problem is identified at its incipient stage, and quantifiable problems can also be expressed on the diagram by noting information such as the size of the problem.

In this way, a 5-why diagram on a single sheet of paper tells which of the production inputs (5 Ms + I) the problem concerns, the reason why the problem has arisen, and the extent of the problem in terms of time, etc. Unlike using a cause-

and-effect diagram, there is no need to write down a lot of complex, disorganized information on factors that are not really causes of the problem. When we are faced with a problem, the only thing we need to do is to identify its one true cause and resolve it. The 5-why diagram contains no extraneous information. If the first set of five "Whys?" does not identify the true cause, we repeat the process for another of the production inputs (5 Ms + I), creating another branch on the diagram. Once the true cause has been identified in the workplace, we stop writing and move on to the next step—that of finding ways to eliminate it.

An important point to note when applying the 5-why technique is to describe the causes in the form of full sentences, not just as single nouns. This is because nouns by themselves are too cryptic and do not adequately express the cause, and a person looking at the diagram for the first time will not really understand what the causes are. Finally, the most important thing of all is to accertain the facts in the workplace when drawing up the diagram.

This concludes Step 3, but the following effective supplementary technique may sometimes be used if necessary. When the question "Why?" is asked and an answer is not immediately forthcoming, it can be helpful to hold a brainstorming session to collect ideas about possible answers. Write the ideas on cards, arrange the cards in groups, take the cards to the workplace, and use them as hints for finding the causes. Another effective approach is to hold a brainstorming session before applying the 5-why technique. This is done to work out the best way of examining the workplace in order to find the true causes of the problem.

Step 4 for Tool No. 3

Once the true causes of the problem have been identified in Step 3, the next step is to rank these causes in terms of the priority that should be given to their elimination. This must be done whenever there is more than one true cause. The purpose of Step 4 is thus to clarify where the countermeasures should be applied. In this step, we decide which of the causes it would be most effective to eliminate and where we should apply the countermeasure in order to solve the problem. Once we have decided this, we show our decision by drawing a circle around the selected cause or causes. Generally speaking, we select only a single cause, but up to a maximum of three can be chosen, if necessary.

Step 5 for Tool No. 3

The final step, once we have identified the single most important true cause, is to decide on an effective countermeasure.

In this way, we use a single diagram to summarize the causes and effects of the problem. The 5-why technique is thus an analytical technique for pinpointing the true causes of a problem.

GO-GO TOOL NO. 4: THE FACT-FINDING (FF) TECHNIQUE

The FF technique is used to perform a quantitative analysis of a problem, and it need not be used at all if all of the true causes of the problem have already been identified using the 5-why technique. The FF technique is used when the true cause of the problem cannot be identified without performing a quantitative analysis. In this sense, the FF technique is a supplementary technique for use with the 5-Why technique.

Step 1 for Tool No. 4

The first thing necessary is to analyze the problem visually and objectively. Identifying a problem in visual terms is effective for speeding up the application of countermeasures. One of the best ways of doing this is to use photographic methods such as fixed-point photography or video analysis. The use of video analysis to obtain an accurate grasp of the existing situation is particularly effective, and this method is described in the following steps.

Step 2 for Tool No. 4

In Step 2, the video films taken in Step 1 are analyzed quantitatively. Manual operations can be subjected to time analysis, and equipment operations can be analyzed by replaying them in slow motion. Various techniques are used at this stage to analyze work procedures.

Specifically, we analyze work processes and perform motion analysis by borrowing the principles of motion economy (when analyzing administrative work, we subject the flow of information to process analysis). Using these various analytical approaches, we clarify which parts of the work are value-adding and which are wasteful, and then proceed to improve the operation based on this information. The techniques mentioned so far are industrial engineering (IE) techniques, but checksheets, histograms, and similar simple statistical techniques may also be used. Whatever technique is used, the purpose of this step is to quantify the problem.

Step 3 for Tool No. 4

Step 3 is for pinpointing exactly where the problem lies. To identify what is actually happening, the current situation is re-enacted by a modeling procedure known as "pantomime simulation" in order to discover where waste is occurring. Slowly re-enacting an operation in this way reveals wasteful motions and other previously-unnoticed sources of waste. By acting out the operation slowly, people realize things about the work they had not understood before—things like,

"This distance is rather long," "It takes a long time to get to this point," "There's a lot of moving up and down," and "Can't that document be obtained without coming right around to here?" If the lines of motion are then drawn on a diagram or on top of a photograph, the waste becomes plain for all to see, and such diagrams also make it easier to explain the problem to others.

Viewing actions in slow motion and discussing them in this way makes it possible to obtain a very accurate grasp of the situation. Slowly re-enacting an operation captured on video or subjected to time analysis enables us to understand sources of waste and put ourselves in the workers' shoes in order to identify what needs to be improved. In the case of equipment, we perform slow-motion analysis. By moving the machinery slowly, we can assess the efficiency of each of its operations.

Step 4 for Tool No. 4

If the above three steps have been properly followed, the problem should now be understood qualitatively. The next stage is to express it quantitatively. It is important to simultaneously use checklists at this stage to note all the points requiring improvement. Merely analyzing the problem and expressing it in the form of graphs or diagrams is pointless; such graphs and diagrams are only meaningful if they also include notes describing the improvements needed.

The problem should be expressed using diagrams such as pie charts, composition diagrams, or graphs showing how the numbers change with time. The VW technique or the 5-why technique may also be reapplied at this point. Whichever method is chosen, it should display the extent of the problem visually. When using the VW technique, information such as the size of the problem, the reason for it occurring, and the points requiring improvement should be noted on the diagram.

Step 5 for Tool No. 4

Since we now know the size of the problem and the reason for its occurrence, the next stage is to clarify exactly what needs to be improved. Once the problem is understood and we know what needs to be improved, we can finally begin to actually solve the problem. To facilitate this, it is important to describe the problem in the form of a full sentence noting what needs to be improved. In other words, we write down the direction we must take in order to solve the problem. By doing this in the form of complete sentences, we clarify our objectives, functions, and roles, and clearly spell out our policy on solving the problem.

This is an important step. We adopt this approach because we want to think first in terms of *objectives* rather than the *means* of achieving those objectives, and then we want to find as many means of achieving the objective as we can and select the best of them. It is usually possible to think of a large number of different ways of achieving a particular objective, and we want to choose the most effective one. It is extremely important in making improvements to express the guidelines for our countermeasures in the form of complete sentences, or at least as noun plus verb, and this is why we also use this approach in the FF technique.

Let me give another example to explain the idea behind this. Imagine that we are considering an operation in which something is fixed into position. For example, in the case of a necktie, the problem might be expressed in the form of the sentence, "Fix the necktie to the shirt." Expressing the problem in this way makes a large number of different solutions possible in addition to the conventional one of using a tiepin or clip; for example, a tie could be fixed to a shirt using a clothes peg, a staple, a bulldog clip, and so on. Many different means of achieving the objective can be found. Someone might even come up with the idea of melting the tie on to the shirt or fixing it on with glue, although neither of these would be very practical. Writing the problem in sentence form enables large numbers of ideas to be brought in from other fields, thereby yielding a wide range of choices from which to select the best plan. Each of the ideas can be examined and the impractical ones discarded; for example, the proposal to fix the tie to the shirt using glue would be dropped because it would spoil the tie and the shirt.

In this way, we start by expressing the problem in sentence form and gathering as many techniques as possible, without restriction, that suit our objective. In this case, the idea of using an ordinary tiepin or tieclip would probably be selected as being the least expensive and most effective. In this way, we clarify our objectives, express the improvement we want to make in sentence form, and select the best plan for achieving the objective.

GO-GO TOOL NO. 5: THE IDEA MAPPING TECHNIQUE

The idea mapping technique is a method of pinning down exactly how we are going to solve the problem (see Figure 3-4). Then, once we have found a good idea, we are finished with the go-go tools and we move on to the implementation stage. The idea mapping technique is a way of sorting out the ideas and identifying the best.

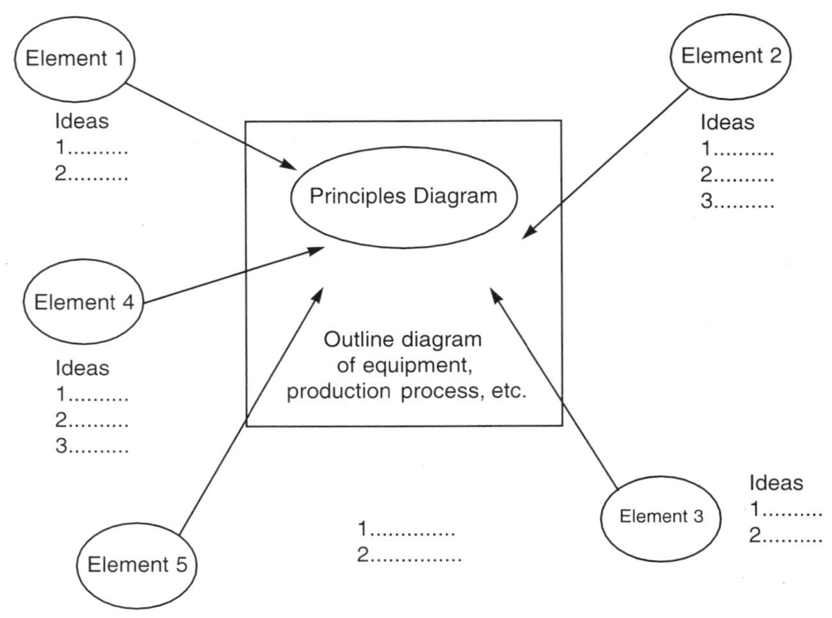

Figure 3-4. Go-Go Tool No. 5: Idea Mapping Technique (Principles Diagram)

Step 1 for Tool No. 5

First, to generate ideas, we examine the work, equipment, or production process we are seeking to improve, and express its principles diagrammatically. Specifically, we make a schematic drawing of the equipment or operation we are looking at and indicate on it exactly where the problems lie. We then divide up what we are going to improve into separate elements and describe these elements, again, in the form of complete sentences.

Step 2 for Tool No. 5

We then hold a brainstorming session to collect ideas on the points we need to take action against, in relation to each of the elements identified in the previous step.

Step 3 for Tool No. 5

In Step 3, we evaluate the proposed improvements using ABC ranking in combination with five-stage numerical ranking. This is done using circular colored adhesive labels (the color of the label representing rank A, B, or C) on which we inscribe a figure representing the value awarded to the proposal on the five-stage ranking scale.

First, the ideas are ranked according to the ABC scheme. Rank A means that the improvement will be easy to effect and the problem will soon be solved. Rank B means that some investigation is probably required before the idea can be implemented. Rank C means that the idea is probably rather difficult to implement (of course, further ranks can be added if required). The ideas are thus ranked in terms of their ease of implementation. This shows us which parts of the overall problem are easily solved.

Next, we rank the ideas on a five-point scale in terms of their degree of importance. We already know which parts of the problem are going to be easy or difficult to solve, and we now need to evaluate them on a scale ranging from absolutely essential to unimportant.

To see how this dual ranking system works, let us consider an idea ranked A-5. We can immediately see that this proposal is easy to implement (A) and, moreover, addresses an issue of vital importance (5). Once we have completed this step, we look for ideas that are both easy to implement and important. Now, we need a way of sorting out the good ideas from the bad ones. To begin, we do a first-stage broad evaluation.

Step 4 for Tool No. 5

In Step 4, we draw up an "idea evaluation" table in order to identify the proposals we want to implement. We note the best proposals on this table and compare them. In Figure 3-5, for example, three proposals (X, Y, and Z) are evaluated. In this example, proposal X was selected for implementation as being the best of the three proposals assessed. This proposal was chosen because it was judged that it would be the easiest to implement and would produce the greatest benefits. This method of comparing and evaluating proposals is far more effective than any amount of abstract discussion.

As noted below the idea evaluation table, we assessed the ideas in terms of how much they would cost to implement, how technically difficult they would be to implement (naturally, this also includes an estimation of the time required to implement them), and how their implementation would impact the people involved. There may be other factors, but, broadly speaking, these are the three basic ones.

Step 5 for Tool No. 5

Once the above steps have been completed, we finally reach the stage of taking the best plan and working out specifically how it will be implemented. In other words, we schedule it. We have finally reached the stage of actually implementing the improvement proposal.

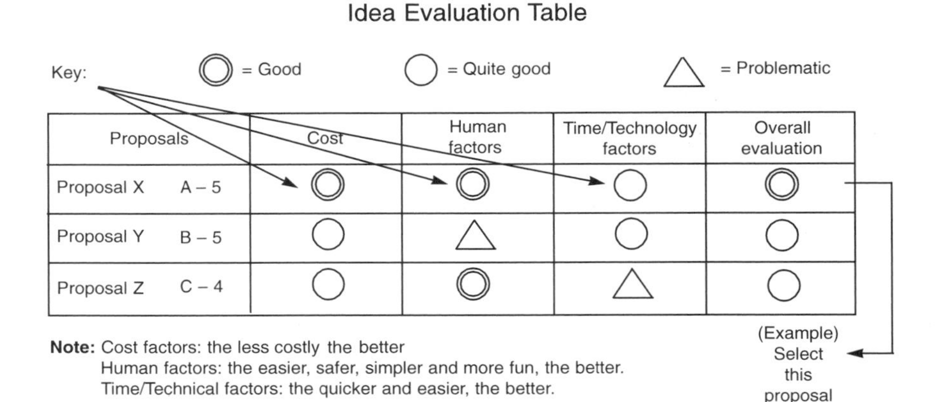

Idea Evaluation Table

Key: ◎ = Good ○ = Quite good △ = Problematic

Proposals		Cost	Human factors	Time/Technology factors	Overall evaluation
Proposal X	A – 5	◎	◎	○	◎
Proposal Y	B – 5	○	△	○	○
Proposal Z	C – 4	○	◎	△	○

Note: Cost factors: the less costly the better
Human factors: the easier, safer, simpler and more fun, the better.
Time/Technical factors: the quicker and easier, the better.

(Example) Select this proposal

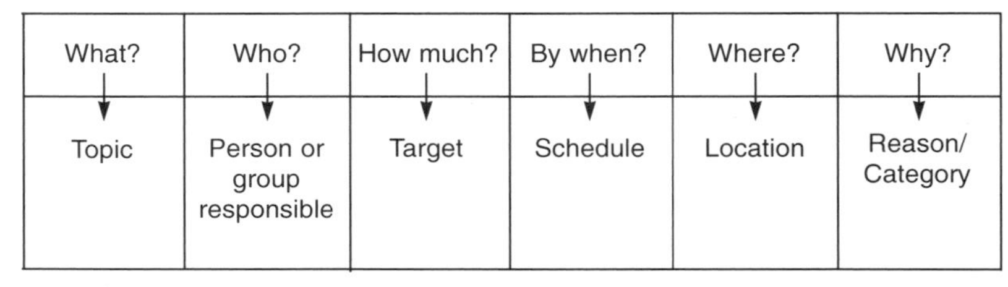

Implementation Planning Table

What?	Who?	How much?	By when?	Where?	Why?
↓ Topic	↓ Person or group responsible	↓ Target	↓ Schedule	↓ Location	↓ Reason/ Category

Figure 3-5. Idea Evaluation and Implementation Planning Tables

To schedule the proposal, we use the 5W1H method. We draw up an implementation planning table and note on it what the proposal is, where it is to be implemented, by when it is to be implemented, why it is to be implemented, how much is to be achieved (the numerical target), and who is responsible for implementing it. Specifically, we note on the table the title of the project, the person or group responsible, the target, the deadline, the location, the reason for implementing it, the benefits of implementing it, and so on. This completes our analysis. Now, all we have to do is actually implement the plan.

The above outline of the go-go tools does not include anything particularly new. We have simply taken the usual methods advocated by the JMA for smoothly and efficiently accomplishing workplace improvements and have arranged them into a set of five steps that small groups should follow as a matter of routine. The following section summarizes the significance of each of the go-go tools.

THE SIGNIFICANCE OF EACH GO-GO TOOL

The Pocket Matrix Technique

The purpose of this tool is to identify the issues and decide which of these the small groups are to address. It is an efficient method of gathering topics and selecting which to tackle.

The VW Technique

The next step is to ask what problems the company is currently facing, how serious these problems are, what the priority is for tackling them, and whether a particular topic should be addressed. The VW technique is a method of prioritizing the topics identified using the pocket matrix technique. The best way of doing this is to arrange the problems in order of their seriousness.

Specifically, we identify the problems and arrange them in order of their seriousness on a block diagram, including both quantifiable and unquantifiable problems. This gives us an indication of which problem or problems to tackle.

The 5-Why Technique

Once we have identified the problems and estimated how serious they are, the next step is to set about solving them. However, there would be no motivation to solve a problem if we did not appreciate the effect of leaving it unsolved. We therefore analyze the effect of leaving the problem unsolved in terms of customer satisfaction, employee satisfaction, and corporate profitability. Once we realize the adverse effects of the problem, we become very keen to solve it. This means that we have to put all our efforts into finding the true causes of the problem.

We do this using the 5-why technique in combination with the 5W1H technique and the three actuals approach, based on the idea that most problems have a single prime cause and a single most effective solution. Of course, there may be instances when several causes come to light in the process of analyzing the problem. When this happens, we compare the causes, assess their relative importance, and devise strategies to deal with those we consider most important. The 5-why technique enables us to identify the causes of problems and move towards their solution.

The FF Technique

In some cases, the seriousness of a problem must again be quantified. Sometimes, frequency analysis is required in order to identify the time factors involved. In these cases, time analysis or video analysis must be conducted in addition to the previous investigations. We therefore gather further quantitative data at the site of the problem to assess the extent of the problems numerically and make it clearer before devising countermeasures.

The Idea Mapping Technique

Having pinpointed the location of the problem, assessed its seriousness, and identified its cause, we now have to generate some specific ideas for solving it. To begin, we express the goal of our proposed improvement in sentence form and gather as many ideas as possible. We make a diagram showing exactly what it is we want to solve.

Once we have collected some suitable potential countermeasures, we evaluate these and select the best. We then examine these in order to choose the one that will yield the quickest solution for the smallest amount of effort.

Once we have decided on the countermeasure we are actually going to implement, we use the 5W1H technique to draw up a schedule. Having worked out who is to do what by when, the only thing left is to take action and follow it up.

TAKING ACTION

So, what can a small group achieve by using the go-go tools and proceeding through the steps described above? Having identified a number of improvement topics that fit in with their company's management needs, they can use the go-go tools to:

1. Understand what the problems are.
2. Find out how they should set about solving them.

The only thing then left is to take action. If the group members cannot take the appropriate action by themselves, they must ask for outside help. In either case, they must take the action required to put their countermeasures into effect. If the procedure described in this chapter is followed, small groups are bound to end up solving the kinds of problems that everyone wants to solve—those whose solution leads directly to improved business performance.

In the first stage of applying the go-go tools (the pocket matrix technique), it is important for managers to participate in the process of identifying topics, since this is the only way in which top-down needs can be effectively combined with bottom-up improvement proposals and action. Also, in operating a small-group program, it is essential for management to ensure that the "What?" and the "Why?" are clearly understood. The topics to be addressed must be identified before the small groups swing into action. Then, once a particular topic has been identified, the next step is to decide how to go about addressing it. Once the "How?" has been identified in this way, the only thing left is to act. This is the crux of the go-go system.

CHAPTER 4

The Application of the Go-Go Tools

This chapter presents two real-life examples of the application of the go-go tools. The first of these is an example of improvements effected at a distribution center by the truck drivers who worked there, facilitated by a JMA consultant. The second shows a method of implementing improvements by using a portable video camera.

EXAMPLE ONE: IMMEDIATE, PRACTICAL APPLICATION

This company was anxious to start improvements right away rather than teach people theoretical techniques without any immediate use. They decided to call it "practical small group training," with the idea that after about two hours of classroom training employees would start to apply what they were learning and tackle some real workplace problems.

This example is summarized in Figure 4-1. First the consultant asked the group to draw up a pocket matrix (go-go tool no. 1) on a flip chart and start writing their concerns on self-adhesive notes. One immediate concern was damage to the truck beds, which was making them unsafe. This and other concerns are shown in Figure 4-2. The next stage was to categorize according to the evaluation criteria, also shown in Figure 4-2. Important problems, marked with the flower symbol, were placed on top of those of lesser importance.

Introducing go-go tool no. 2, the VW diagram, the consultant asked one of the managers to help the group assign values to the problems that were identified as important (see Figure 4.3 on page 50). Then Pareto analysis showed how serious the problem of damage to truck beds was.

Deciding to tackle this problem, the group then used go-go tool no. 3, the 5-why technique. To do this, they first wrote their problem statement in a box in the center of a piece of paper (see Figure 4-4 on page 52). Looking at the consequences of not correcting the problem first, the group soon worked out that it would have a direct effect on their reputation as a company, the safety of the drivers, and their ability to earn more.

The group then brainstormed the causes of the problem. Their first effort focused on the drinking habits of one of the drivers, which were held indirectly responsible for the latest incident. The consultant encouraged the group to move beyond this explanation and look at how the damage was physically caused, and how this could be prevented.

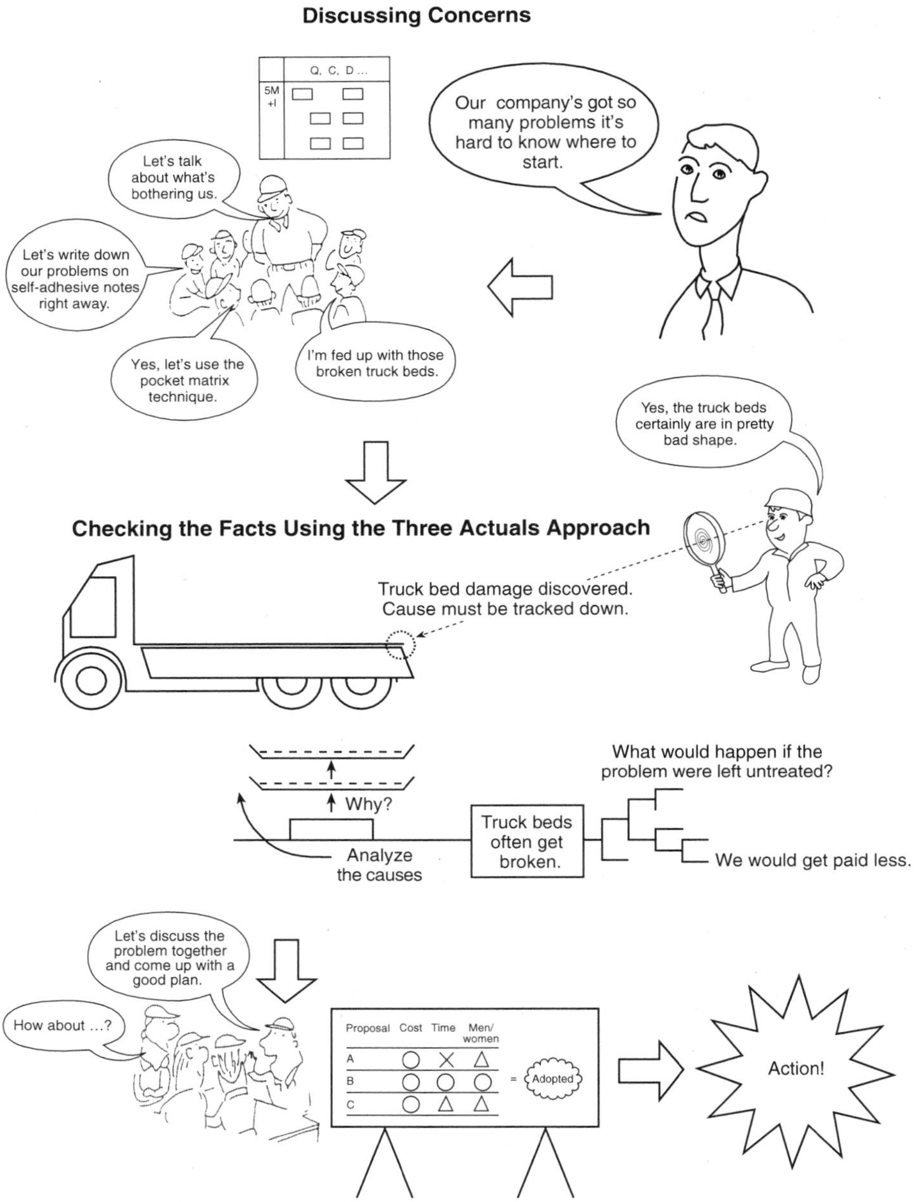

Figure 4-1. Practical Small-Group Training for Real Workplace Problems

Evaluation Criteria

- ❀: Can be achieved quickly and will definitely be beneficial. Should be implemented quickly.
- ◎: Can be achieved relatively quickly, and some benefits can be expected. Best implemented soon.
- O: Appears likely to produce some benefits, but will probably be difficult in practice. Should be implemented soon if possible.
- Δ: Will take a long time to implement and will only produce slight benefits.
- ×: No benefits can be expected. Not worth implementing.

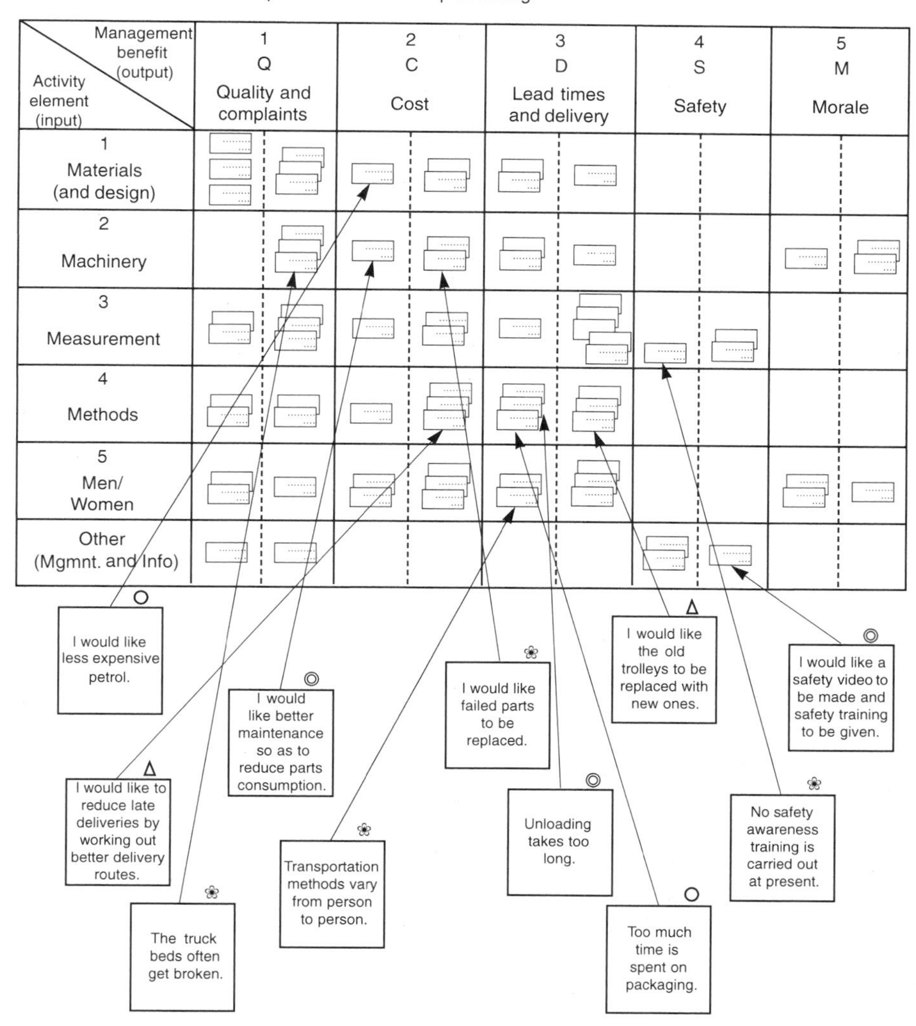

Figure 4-2. Pocket Matrix—Concerns of Truck Drivers

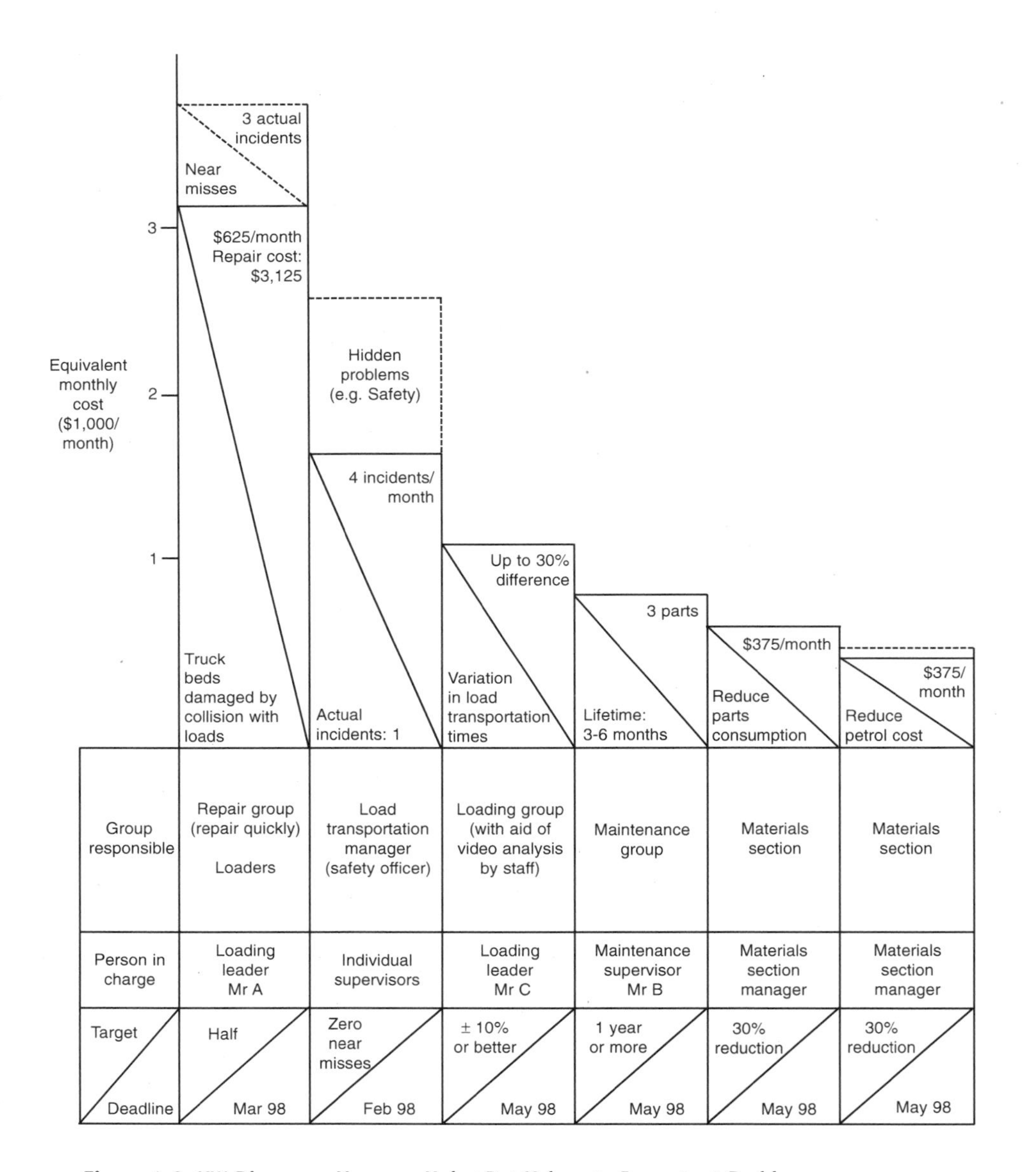

Figure 4-3. VW Diagram—Manager Helps Put Values to Important Problems

In order to do this, they would have to gather more facts by employing go-go tool no. 4, the FF technique. To do this they made a video recording of the operation of loading the trucks, and analyzed it in slow motion (see Figure 4-5).

They had discovered that, if the crane was moved forward at a speed of 30 m/min and stopped just behind the truck bed, the load would swing through a 1-meter arc. The slow-motion replay of the video showed that the loader had to operate the crane carefully in order to prevent this. Although this was a fairly simple method of analyzing the situation, it clarified the problem effectively. The group members had talked vaguely about solving the problem by standardizing the method of operating the crane, but when the consultant arrived, they decided to write down their ideas and discuss them.

Having written down a number of ideas, they decided to go to the work site to see if these ideas would actually work in practice. While discussing their ideas, they drew up the principles diagram (idea mapping technique) shown in Figure 4-6 on page 54. They divided their ideas into five areas, wrote these down on the principles diagram in sentence form, wrote down the purpose of the improvement (also in sentence form), then wrote their ideas on self-adhesive notes and fixed these to the diagram.

Everybody was extremely eager to try out their ideas. In the past, the company had simply taken the advice of experts on what to do; working out their own ideas and deciding which ones to put into practice was a completely new experience for the drivers. Before, if no consultant had been available to give advice, the company's senior managers and technical staff had decided what to do. Ordinary employees could make suggestions, but they were often ignored.

When the consultant returned for his final visit, the group showed him the evaluation table they had drawn up (see Figure 4-7 on page 55). They explained that they had narrowed it down to two ways of stopping the load from swinging. Since the crane could not be lowered, they thought about putting a swing damper on the cable. They tried this using a mock-up and it was successful. Then, to prevent the truck beds from getting broken, they decided they needed to install a shock guard, which they also tested using cardboard. This worked even for different heights of trucks. The group was now going to meet after work to put their plan into action.

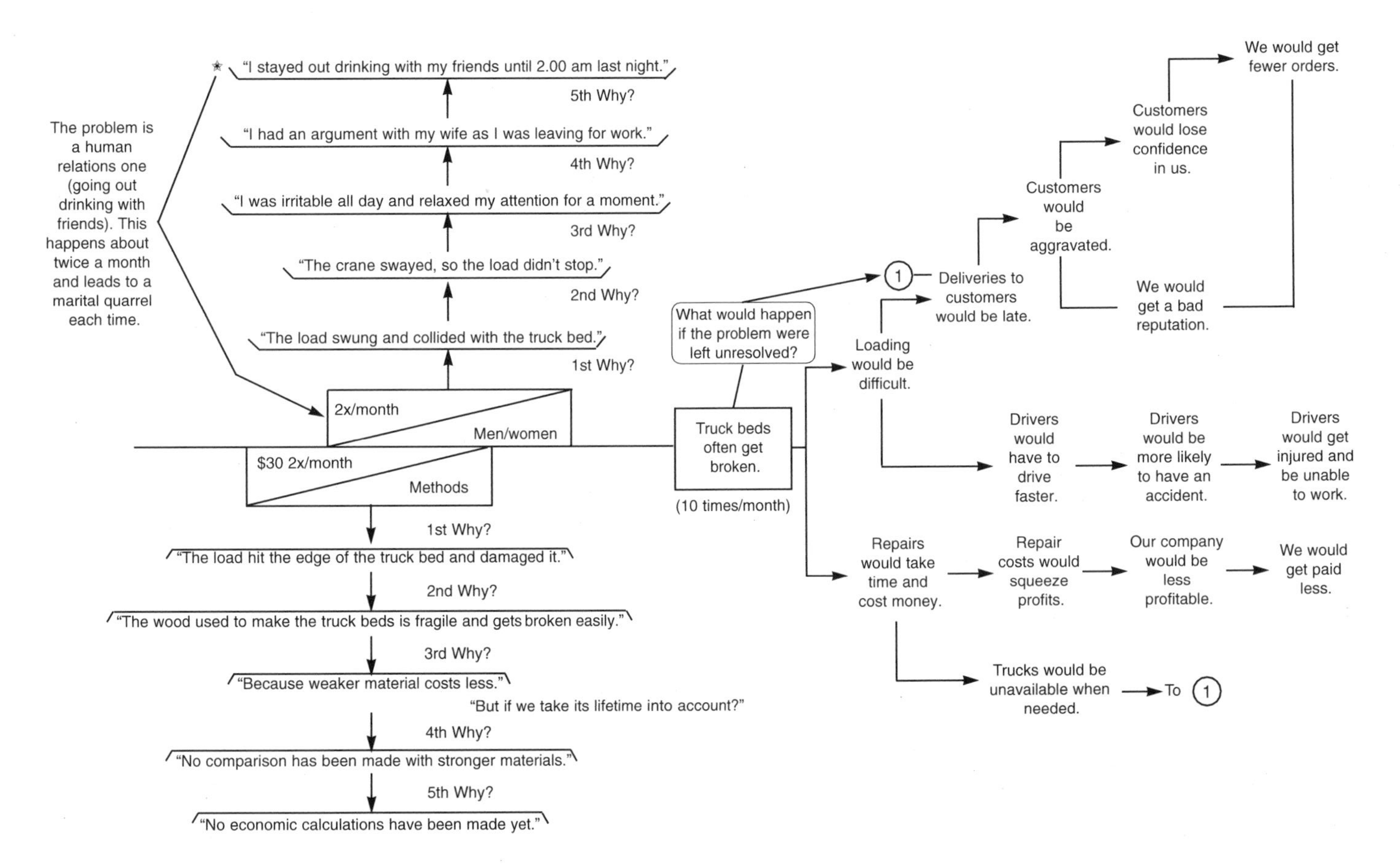

Figure 4-4. 5-Why Technique—Tackling the Most Serious Problems

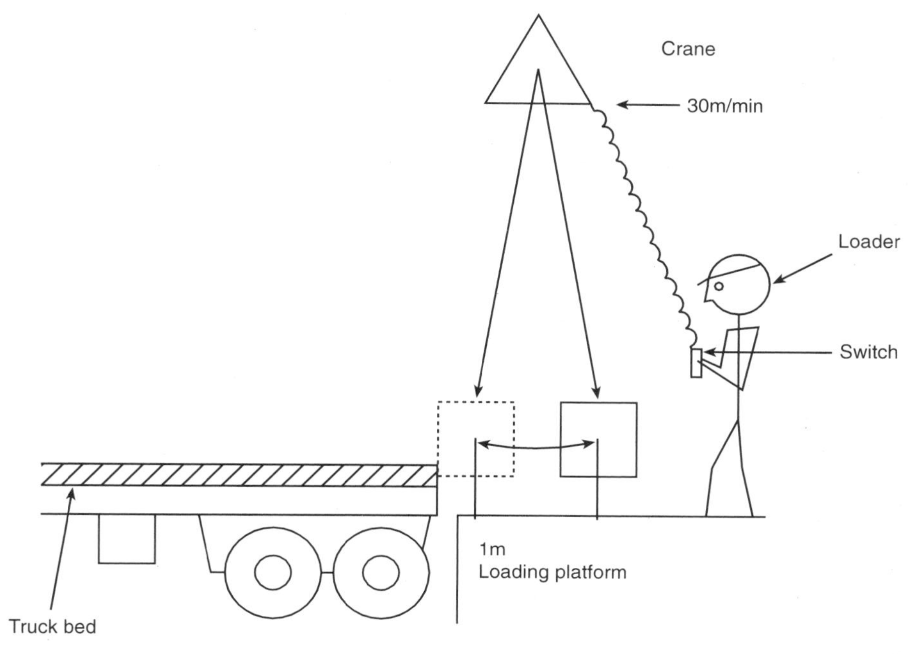

Figure 4-5. FF Technique—Video Recording and Analyzing the Loading of Trucks

Features of the Go-Go Tools As Applied to Example 1

1. The small groups were aware of the management issues, and they started off by skillfully linking their own concerns with the topics they were required to address.
2. They used the go-go tools systematically with the aim of achieving good results with the minimum labor input. In particular, they followed the principle of the three actuals when tracking down the causes of problems. Because of this, their approach was very practical and was focused on achieving useful results rather than on drawing up charts and tables. This meant that the improvement cycle was speeded up.
3. Because they had looked at the problem from all angles and had collected a large number of ideas for improvement, whenever they had finished one topic they were able to move quickly on to the next. Also, because the small groups were formed around topics rather than around the members of particular workplaces, the groups contained a good mix of people and their members were able to learn from each other while solving the problems. This helped dissolve barriers between different workplaces, and everybody felt that they had developed their abilities to perceive and analyze problems.

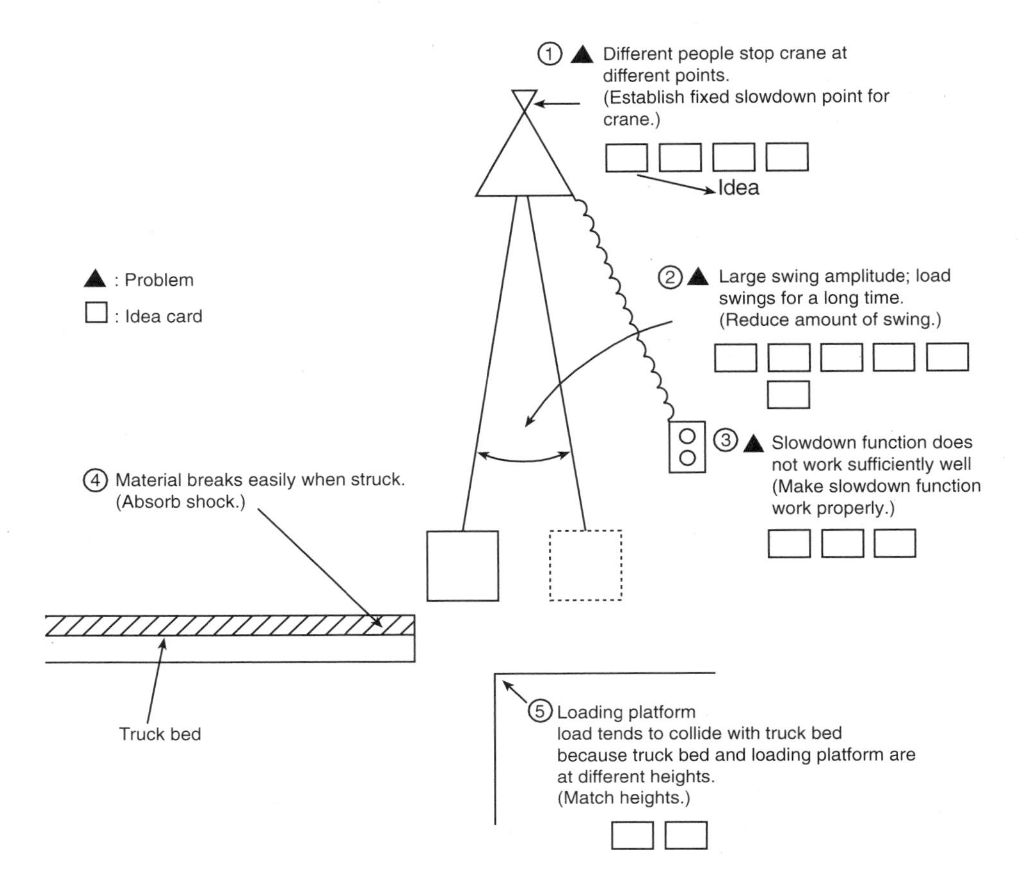

Figure 4-6. Principles Diagram—Dividing the Teams' Ideas into Five Areas

EXAMPLE TWO: IMPLEMENTING IMPROVEMENTS USING PORTABLE VIDEO EQUIPMENT

The previous section explained the use of the go-go tools by means of a case study. I would now like to discuss a particular method of implementing improvements using portable video cameras. This technique has recently been developed by the JMA and is beginning to be used by large numbers of companies as a highly effective tool for accomplishing rapid improvements in the workplace.

This technique is one of the methods used in go-go tool no. 4, the FF technique. The example described here illustrates the use of the Sharp Viewcam together with a Sony video printer.

Proposal	Cost factors	Human factors	Technical factors	Overall evaluation
Proposal 1: Control speed of crane (improve circuit and replace motor.)	Δ Would cost $3,750.	◎ Would remove worry from loading operation.	○ Feasible, since company A already does it.	Δ Implement at appropriate time; for now, make a stop mark on the underside of the crane track.
Proposal 2: Install a shock guard. Erect guard when loading Spring Loading platform Truck bed Hinge type	◎ We can easily make a cheap guard ourselves.	○ It would take time to erect the guard each time.	◎ We could do it very soon.	◎ Start today, and fit guards little by little each month.
Proposal 3: Install swing damper on crane. Guide Cable	◎ Inexpensive and simple.	◎ Good from safety viewpoint too.	◎ Could be done soon—we could make it ourselves.	◎ Do immediately.

What?	Who?	By when?	Where?	Why?	Follow up?
Proposal 2	Mr B	May 1998	• Truck beds • Loading platform	To protect truck beds from impact	
Proposal 3	Mr C and equipment department	May 1998	Cranes	Including safety	

Figure 4-7. Idea evaluation and Planning Table—Showing Two Best Ways of Stopping the Load from Swinging

Triple-Speed Kaizen—How to Solve Problems and Effect Improvements on the Spot at Three Times the Usual Speed

The aim of effecting improvements using a Viewcam and video printer is to speed up the improvement cycle by photographing the work on the spot, discussing it right there and then, and coming up with improvement ideas right away by physically moving around instead of just sitting and talking. At the present time, the procedure at most companies is to take a video film in the workplace (step 1), return to the office or classroom, look at the video, discuss ways of improving the operation, and put the ideas on paper or on a blackboard (step 2), then return to the workplace to check the suitability of the improvement proposal (step 3).

From now on, however, improvements will increasingly be made as shown in Table 4-1, by going to the workplace and devising improvement plans there through a process of reenactment and simulation. The above three steps are condensed into one single step in this way, and this is why the new approach is known as triple-speed kaizen (kaizen means continuous improvement using the plan-do-check-act cycle).

As illustrated in Figure 4-8, the work is reenacted in or near the actual workplace, and the re-enactment is filmed. The film is then analyzed with the aim of identifying better ways of carrying out the operation. This figure shows an actual improvement project carried out at an American company that manufactures high-grade document binders. The improvement team first filmed the actual operation in the workplace, then reenacted it and devised ways of improving it. The photographs show that, in the original operation, adhesive is applied to a number of pieces of card to be used as the binder's stiffeners, and these are then spread out on a larger card to dry. The team decided to improve this operation by using a rotary-type holder instead of the large piece of card. They made a model of the rotary-type holder, simulated the improved operation, and filmed the simulation. The effect of the improvement could then be assessed by comparing the film of the operation before and after improvement.

Using the Viewcam's timer, it was found that a 50 percent time saving had been made. The video printer was then used to obtain still photographs of the improved operation, and these were used as standards, obviating the need for further documentation. The proposal was immediately discussed in the workplace and put into action (the proposal was made in November and implemented in January).

This real-life example shows just how effective the Viewcam and video printer can be for making improvements. I sincerely hope that this fast, easy, and practical improvement technique will be used extensively in workplaces in the future.

Table 4-1. Devising Improvement Plans Through Re-enactment and Simulation

Item	Conventional Approach	Triple-Speed Kaizen
Discussion Style	1. Take video of operation in workplace. 2. Discuss video in office or classroom. 3. Return to workplace and check improvement ideas (at least three steps).	1. Take video of operation in workplace. Perform simulation to identify existing situation, then devise and check improvement idea and prepare materials.
Documentation	1. Video analysis sheets. 2. Improvement notes and drawings. 3. Teaching materials and written standards.	1. Take video of operation before and after improvement (use Viewcam's timer). 2. Compare situation before and after improvement and standardize by means of still photographs obtained from video with video printer. Note key points only in writing (no need to prepare lengthy documents).
Use of techniques	1. Take video of operation and check using IE or QC analysis sheets. 2. Sometimes training is divorced from actual use of techniques. 3. Either IE or QC techniques are used, not both.	1. Note and utilize IE, QC, and VE ideas. 2. Improve operation while checking by simulation. 3. Use slow-motion analysis and other techniques to simultaneously check improvement ideas in workplace.
Operator involvement	1. Focuses mainly on preparing materials to explain situation before and after improvements. 2. Investigations into safety of improvement are begun after improvement has been announced. 3. Workers only become involved once improvement proposal has been formulated.	1. Improvement is explained by means of simulation; improvement is put into practice and videoed. 2. Improvement is checked in workplace and implemented with workers' assistance.
Speed	1. Take video. 2. Discuss improvement. 3. Prepare documentation and try to make everything perfect.	1. Immediately put improvement into effect and proceed to next round of improvement. 2. Minimize documentation.

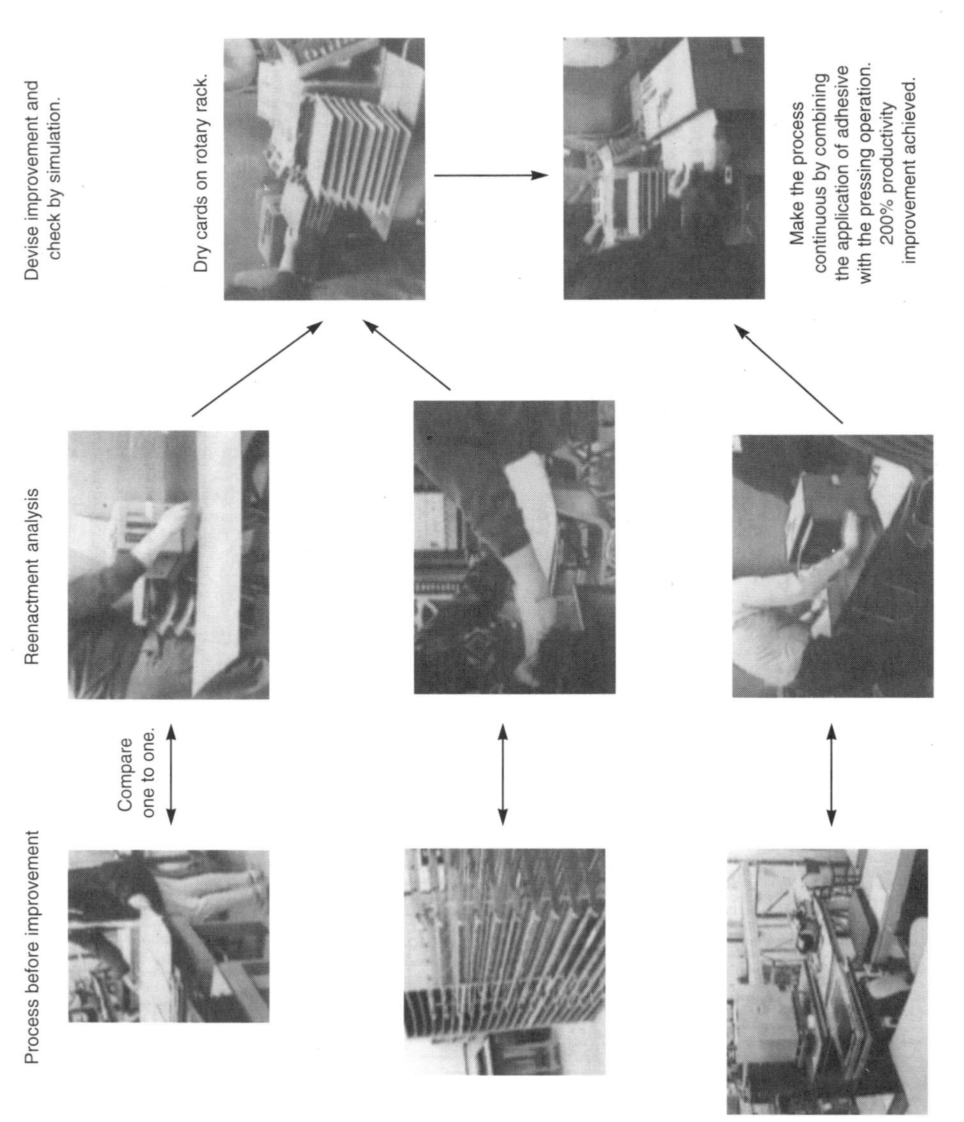

Figure 4-8. Filming the Work Re-enactment in or Near the Actual Workplace

CHAPTER 5

The 5-Sheet Presentation Technique

If the advice given in chapters three and four is followed, small groups are bound to start producing a variety of significant improvement results. This section describes a simple 5-sheet presentation method I have devised to enable groups to present their achievements effectively to each other and thereby raise their general level of technical ability.

I settled on a limit of five sheets because one overhead projector (OHP) transparency usually takes about three minutes to explain, so five transparencies will take 15 minutes. Some of the sheets will probably be simpler to explain than others, so I estimate that in practice the actual presentation time will generally be about 13 minutes 50 seconds. The essence of the method is to use a simple format and no more than five or so OHP sheets to present information with real and substantial content. (Extra figures and tables may be included in addition to the basic five sheets, provided that these are self-explanatory.)

Before proceeding to the main discussion, I would like to recount briefly how I came up with the idea for the 5-sheet presentation method. It started when I paid a visit to a factory to hear a presentation on energy-saving by a recently-promoted site foreman. Seeing me, he came up to me and asked, "How many minutes should I talk?" The company had a long history of quality improvement, and its small-group activities were now being redirected towards the question of energy-saving. The group under this man had produced outstanding results, and he had received his promotion because of this. I immediately inquired why he had asked this question. He replied, "I've studied quality improvement pretty thoroughly, you know. You have to ask the customers to listen to you so that you can provide them with the most valuable quality information, and you have to get the information across to them in a set amount of time. The same thing applies to presentations. We must take the small-group methods and activities we've used up till now in studying QC and apply them to the subject of information and communication. If we don't, then we won't be making good use of what we've learned, will we?"

Hearing this, I was really impressed by this man's excellent attitude. He is still leading group activities, and he always specifies a time limit for his presentations and condenses his explanation about what has been achieved into that time. Moreover, he is always thinking about how he can make the content of his presentations useful to his audience. I was full of admiration for what he was doing. Rather than seeking self-satisfaction by being praised for his results, he was striving to take the things that he learns during his everyday work and put them to practical use.

Applying this approach means presenting information in such a way that it gets through to the audience virtually immediately. In an office setting, this would mean something like distributing a single hand-out or showing a single OHP slide and explaining it in 10 minutes or less. At a larger meeting at which tables of data are presented, a 5-sheet, 15-minute presentation would seem appropriate. Since my visit, I have put a lot of thought into this subject and have carried out various tests. These tests have confirmed the effectiveness of the 5-sheet presentation method introduced here, which I have been using for a considerable time now. A simple description of the method is given below.

AN APPROACH TO CREATING PRESENTATION MATERIAL

The method described here enables the presenter to skillfully organize the items to be included in the presentation. The idea is to provide exactly the right information at exactly the right time. Information costs money, so it is expensive and wasteful, as well as pointless, to present reams of meaningless facts. We must ask ourselves what the really useful information is, and seek to provide only the information that the people listening to the presentation, especially those who are going to take the contents of the presentation and further develop them, need to know. This is a vital point to check at this initial stage.

Top managers require information for making policy decisions and other judgements, or for assessing the results of their actions. For middle managers, the emphasis should be on breaking down and simplifying top-management policy, pointing out unusual problems, working out what middle managers should be doing together with their staff, deciding priorities, looking at how to narrow down the focus, and so on.

Front-line workers, on the other hand, need more specific, practical information, but this is not the kind of information that should be provided at presentations. Presentations should be geared to the information requirements of top and middle managers. Nonmanagerial people may listen to the presentations and gain some technical hints, but more in-depth discussion should be left until after the presentation, when people can be invited to the presenter's workplace to hear about the activities in more detail. If the nonmanagerial people accept this first invitation, then as far as they are concerned the presentation has done its job. Companies should reject the notion that presentations must explain anything and everything.

BASIC STEPS FOR COMPILING INFORMATION

There are five basic steps to take when compiling information for presentations.

Step one: Gather information

Prepare some cards and write down anything you might want to include in the presentation. This will create a large pool of key words. Next classify these broadly into groups.

Step two: Assign a priority weighting to each card

Indicate the order of importance by marking the cards with a double circle, single circle, or triangle, or with a red, yellow, or blue dot. It is immaterial what scheme you choose; the idea is to highlight the items you want your presentation to emphasize. These cards will be used as a basis for preparing the 5-sheet presentation materials. Line the cards up, work out an overall structure for the presentation, and divide the cards into five groups.

Step three: Consider how you will organize the presentation as a whole

Decide what kind of OHP sheets you will prepare. What theme will convince the audience of your success, convey the effort required to achieve it, and be capable of being demonstrated by data? To make the important points easy to understand, think up an image to represent each of the five groups of cards.

Step four: Decide exactly what you are going to say

Let us say the average person talks at a rate of about 150 words a minute, about the same speed as a broadcast announcer talking normally. A good way of making the length of your talk flexible is to divide the contents up according to a three-stage scheme using different-sized alphabet characters. Use the largest characters for the information you want to stress the most, medium-sized characters for key words representing the main points of the talk, and small characters for supplementary information, which you can leave out if you run out of time. With this, the structure of the presentation is virtually complete.

Step five: Add the finishing touches

The overall shape of the presentation has now been decided and the only thing left to do is to finish it off and ensure that the OHP transparencies are attractive and well-organized. Use eye-catching pictures or even cartoons to help make your presentation dynamic and interesting.

If you follow these five steps, you will have worked out the theme, the points to emphasize, and the contents of your talk, and you will simultaneously have been brushing up your presentation techniques. You can also make any necessary alterations as you go along. Without writing the presentation out in full, you will have enough material to talk about ad lib, and you should also have no worries over time management, as you can adjust the length of your talk by including or leaving out key word material. This system enables you to make your presentation exactly the right length.

HINTS FOR DRAFTING THE THEME

It is important to develop a knack of creating a good theme, or "story," that will hold the presentation together and keep it interesting. The method I will describe should be used in Step 3 of the five-step process just explained. It is best to visualize the presentation as unfolding in four stages, with the theme being introduced, and then in turn developed, illustrated, and concluded.

Start by writing down the purpose of the improvement project, together with an outline description of your job. This constitutes the introduction to the presentation. Next, develop the theme by noting why the improvement was needed and explaining its background and significance. Write down the ways in which solving the problems were expected to improve the company's business performance. Be sure to explain this clearly and in specific terms.

Then, to illustrate the theme, use actual facts and examples relating to improvements or problems to convey a clear picture of the situation before and after improvement. What were the major topics and problems? What policy was adopted in addressing them? What difficulties were encountered? How was the improvement actually implemented? This is probably the most important part of the presentation. Demonstrate the quality of your improvement efforts by placing the emphasis here, and even showing some data to illustrate what you have achieved.

The final stage is the conclusion, which should include your ideas about what you would like to see happen in the future. Each improvement is no more than a milestone on an endless journey, so a good presentation should always include a summary of the problems that remain to be solved, as well as a description of the results achieved so far. The idea of this section is to stimulate discussion and obtain opinions from as many people as possible on how to proceed with the next round of improvement. Figure 5-1 illustrates the type of OHP transparencies you might want to prepare to illustrate your talk.

HOW TO BE EFFECTIVE

The audience will understand what you are saying more easily if they hear your conclusion right at the beginning. So the best approach for an effective presen-

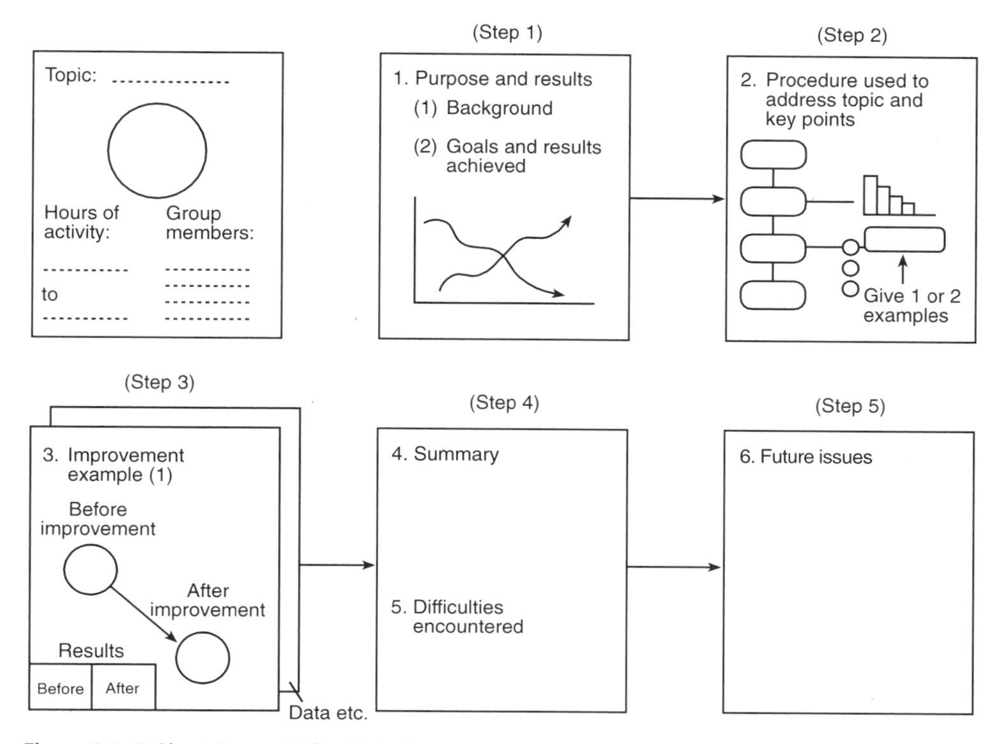

Figure 5-1. 5-Sheet Presentation Technique

tation is to start by stating very briefly the purpose and results of the improvement. Follow this by giving some background information to explain why this particular topic was tackled, and then describe the results achieved in more detail. This is the stage at which you explain what you did and describe how you applied the problem-solving approaches of interest to your audience. Hearing this, the people attending the presentation will appreciate your skillful approach and be drawn into your talk.

A PROCEDURE FOR ADDRESSING A TOPIC

The best way to tackle the topic is to use a single flowchart supported by data (see Step 2 of Figure 5-1). Another good way is to write down a list of key points as used in the go-go tools procedure and highlight what was done at each step by means of pictures on the right of the list. The aim here is to get the listener to understand how the topic was addressed and demonstrate that it was tackled methodically, by a well-thought-out procedure. There is no need for a lengthy explanation here.

In Step 3 of Figure 5-1, one of the key points in the list is described in detail to give an actual example of one of the improvements effected. In my view, this should be the high point of the presentation. Here, you use diagrams and tables to describe the situation before and after the improvement and discuss what was done to effect it. Then, if necessary, you supply data to describe the effort that was applied. It is particularly important to note the results in large characters in a prominent position on the slide.

Step 4 of Table 5-1 summarizes the conclusions reached. What was the outcome vis-a-vis the initial objective? At the start of the presentation, the conclusion was stated very briefly, and here a little more detail is given about the results achieved. Particularly in presentations on small-group activities, it is effective to describe the difficulties the group members encountered and overcame at this point, since their colleagues in the audience will be able to relate to this and appreciate the team's efforts.

Finally, describe your vision for the future, indicating the new directions that the improvement drive will take. Small-group improvement means upgrading step by step, and taking on an endless series of small improvement topics at a rapid pace, so it is important to mention the topics to be addressed next.

Although I have stated as a guideline that you should use five sheets for the presentation, this does not mean that you must spend three minutes talking about each. You can vary the emphasis, so that a less important part of the presentation may only take one minute, while a more important one may be given four to five minutes. The various parts should be appropriately balanced within the overall 15-minute time limit. The key to delivering a good talk is to check the OHP sheets while you are talking, since these contain the key points. Then, even if you forget what you want to say, you can convey the main thrust of your presentation by focusing on the important points written in large characters.

Another secret of giving a good presentation is to state a precise goal each time you show an OHP sheet and describe clearly the techniques used to achieve that goal. In other words, specify the objectives and the means employed to attain them. It is important to list the key points in large letters, since this enables you to make your points in an easy-to-understand way, backed up by the diagrams and tables.

SOME SPEAKING TECHNIQUES

Point to the item you are talking about

You can of course read from notes, but this is liable to make the audience think that you have forgotten what you did, or that you are not really involved with them. By pointing to what you are talking about and checking the items off one by one to remind yourself of what you want to say, and adapting your remarks to suit the response of the audience, you will be able to communicate your achievements and the spirit behind them.

Raise your voice for the important points

It may sometimes be necessary to repeat an important point, and some speakers will pause for a moment to gauge the audience reaction before continuing. Although advanced techniques like these are used by some presenters, I believe important points can be stressed simply by raising your voice and repeating what you have just said, adding a comment to indicate that the point is a particularly important one. This will have the same effect as a dramatic pause. There is no need to go overboard; simply repeat the key point a little louder and in a deliberate, relaxed way. If the presentation is over-rehearsed and becomes a theatrical performance, it may entertain the audience but will not necessarily give them a real grasp of the subject.

Follow the presentation's overall theme

Do this by bearing in mind the meaning of each OHP slide, and by keeping your remarks pertinent.

The three techniques described above will enable you to produce a confident and well-crafted presentation with only a few rehearsals.

MAKING THE PRESENTATION DYNAMIC AND INTERESTING

Your presentation will be enhanced if the audience asks questions, and it might even be a good idea to plant one or two people in the audience to ask some. After a 15-minute presentation such as has been described here, some members of the audience are bound to have questions. A question-and-answer session provides an opportunity to give an extended explanation of what was said in the presentation and also to give more detail on points that could not be fitted into the allotted time. A general question might be, "What kinds of difficulties did you encounter?" but the best questions will be those based on the contents of the OHP slides that will elicit information on the problems met with and the secrets behind the improvements effected.

To get ready for questions, prepare photographs or OHP slides showing how the go-go tools outlined in Chapter 3 were used. Although this is raw, unpolished data, it is fine to present it and explain the details in response to questions. Since it shows what was actually done, it will impress people with what was achieved. There is no need to draw up a lot of special material to respond to questions, just as it is unnecessary to prepare yourself for various imaginary questions. If you are asked about a detailed point, and the questioner is from the same company and your workplace is nearby, you can invite the questioner to visit your workplace. If this is not possible, someone involved in leading the small group can reply. In either case, this should be sufficient to give the questioner a satisfactory response.

It is important to avoid holding presentations for their own sake. The whole point of presentations is to convey an accurate impression of the efforts exerted and the results achieved. The aim of a presentation is to communicate effective improvement techniques to large numbers of people. The worst kind of presentation is one that seeks simply to be praised for being smooth and skillful because of impressive handouts and beautiful OHP transparencies. If this approach is adopted, the whole purpose of the presentation will be lost and the presentation will be no more than a dramatic spectacle.

The methods described above are the essential tools for producing effective presentations. I hope that all those involved in coaching small groups will introduce them to all small-group members, so that the time spent on small-group activities does not end up being used for preparing technically-accomplished presentations that make an entertaining exhibition but are devoid of any real content.

These methods can also be applied to workplace announcements and public conferences, as well as to presentations. At some companies, OHP slides used in presentations are later displayed on the walls of the staff lunchroom, and are of great interest to people who were unable to attend the presentation or could not see the OHP slides very well due to the size of the hall. Presentation meetings are not the only means of exchanging technology and reporting results. I very much hope that methods like these, which achieve their objectives simply and directly, will be more widely used and further developed in the future. (One future direction might be the use of the "single-sheet" method for delivering skillful and efficient presentations. This can be achieved by a slight adaptation of Figure 5-1).

CHAPTER 6

TP Management and Small-Group Activities

This chapter focuses on the basic principles of small-group activities used by Japanese companies. It describes recent developments in TP management and their relationship to small-group activities, a TP deployment diagram, a five-step evaluation scale, basic requirements for running a program of small-group activities, and the reason why small-group activities are necessary.

The management approach known as TP management (total productivity management) was proposed over ten years ago by the JMA, and TP prizes are now presented every year to organizations that have set an example for the rest of Japanese industry to follow by developing new management systems and proving them in practice. Over thirty companies have now won prizes. Companies that introduce the TP management approach find that their management efficiency rapidly improves, their overall quality, cost, and delivery performance gets better, and they are able to organize their employees more effectively and direct them towards solving the company's problems while simultaneously improving employee satisfaction. For all these reasons, TP management is currently enjoying an excellent reputation.

THE TP DEPLOYMENT DIAGRAM

Figure 6-1 is a schematic diagram illustrating the TP management approach. This kind of diagram is called a TP deployment diagram or TP matrix. The TP deployment diagram is an effective visual application of the plan-do-check management cycle. It works by structuring the connections between macro-objectives and micro-objectives, thus enabling us to calculate the cumulative total of the micro-results achieved and compare this total with the goals of the macro-objectives. The following paragraphs give a simple explanation of the principles of TP Management based on this diagram.

In Figure 6-1, top management's overriding objective is to increase the company's sales. The first thing in applying TP management is to take the overarching theme of increasing sales and decide which overall objectives must be established in order to achieve this. This is done from a customer-focused perspective, taking into account all the company's circumstances.

The company has decided that it needs to increase the quality of its products and reduce its manufacturing lead times in order to attain its overall objective of increasing sales. It therefore works out exactly how much each operating site

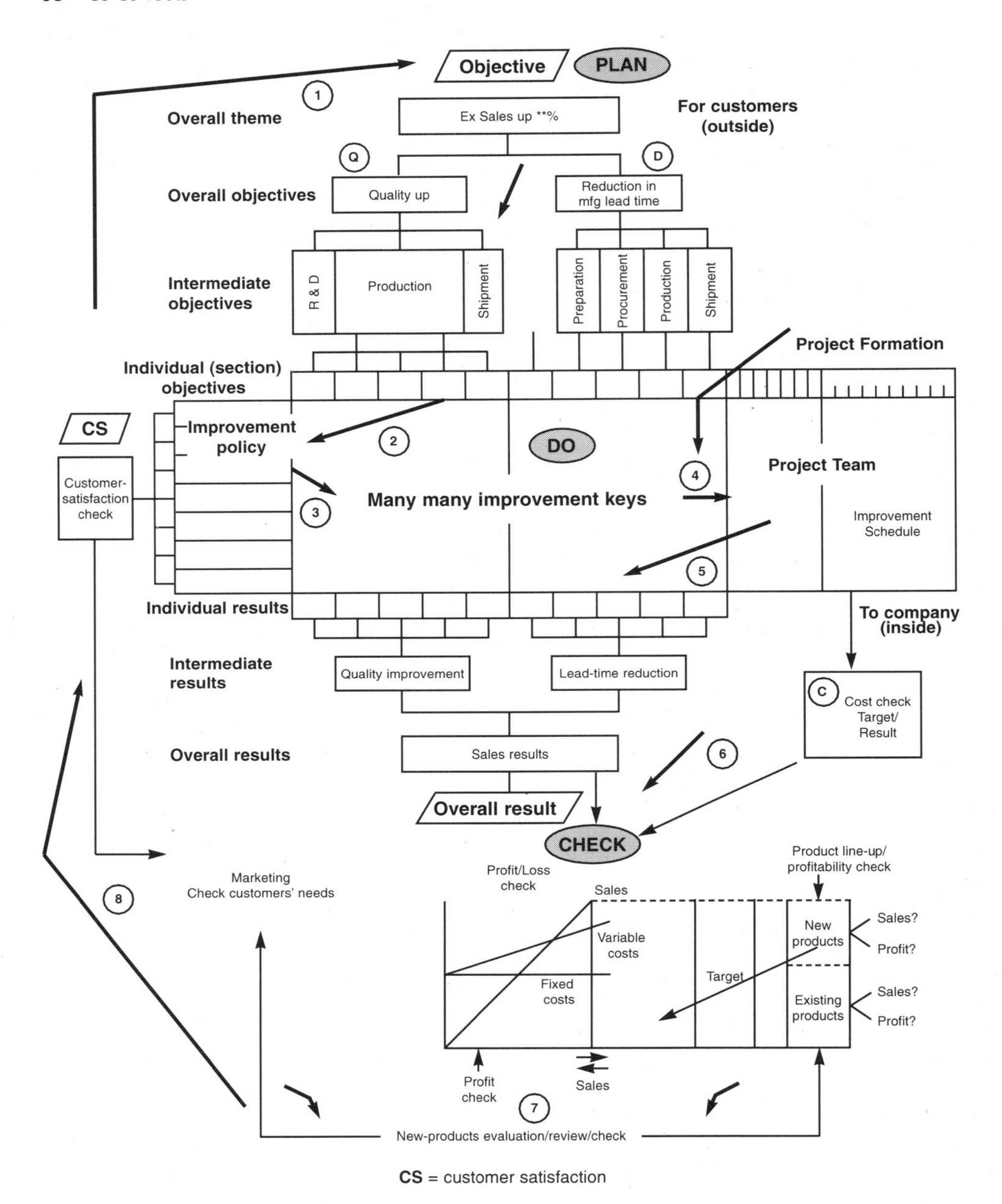

Figure 6-1. TP Deployment Diagram

must increase the quality of its products, and then breaks this target down by department and individual workplace to yield intermediate and individual objectives. Numerical targets are assigned to each workplace, and a mechanism is set up for calculating how much each individual objective contributes to the achievement of the overall objective. This process is termed "objective deployment." In this example, it is carried out separately for the quality improvement overall objective (on the left of the diagram) and the manufacturing lead-time reduction objective (on the right of the diagram).

The first step in applying TP management is thus to proceed from top to bottom, as indicated by the arrows on the diagram, breaking down the overall objectives step by step into individual objectives for specific workplaces (or for specific products). When this has been done, each workplace should know exactly what proportion of the overall objective it is responsible for achieving.

The second step is to scrutinize what customers and competitors are doing and see where they are heading in order to recast the company's own strategies, decide what it ought to be doing, and work out clear guidelines for improving its operations. To begin, we put ourselves in our customers' shoes and try to decide what kinds of strategies we ought to adopt. We perform marketing analyses to check whether or not we are really meeting our customers' needs and what we must do if we are not. As a result of this activity, we should come up with a set of guidelines to help us decide what to do in terms of increasing customer satisfaction. For example, we might decide what kinds of quality improvement are needed from the customer's standpoint, and these would then become items on the list of strategies at the left of the diagram.

Having broken down the overall objectives and decided on strategy guidelines by analyzing customers' needs and trends in competitors' activities, the next step is to decide which specific strategies we are actually going to implement. To do this, we have to consider a wide range of possible strategies while keeping in mind our processes and products and the balance between quality, cost, and delivery. At this point, we make an effort to gather as many ideas as possible from front-line employees. Having decided on the strategies in this way, we prepare to take action by working out ways of breaking down interdepartmental barriers and really coming to grips with the problems.

Having collected a large number of specific improvement strategies, we then prioritize them in terms of their degree of importance, the contribution they will make to the achievement of the overall objectives, and their effect on customer satisfaction. Since the ideas for the improvement strategies have all come from the front-line employees (that is, from the bottom up), they consist of the kinds of topics that will benefit those employees while helping to achieve the overall objectives. The organization and schedule for implementing the selected strategies are noted on the right of the TP deployment diagram. In this way, we plot all the information required for taking action on a single sheet of paper, combining top-down objectives with bottom-up strategies and positioning small-group activities as part of people's normal duties. Preparing this diagram is, in fact, the first step in running an efficient, consistently-directed program of small-group activities leading directly to improved bottom-line results.

In this way, a system of small-group activities designed to achieve the objectives is set up for each strategy or topic. However, in this example, the internally-oriented indicator of cost, that is, cost reduction, is missing. In this case, it is therefore necessary to check each strategy from the cost viewpoint using a separate framework called the profit/loss check, illustrated under the TP matrix. Everything is checked in the same way as it would be if we were drawing up a budget. The purpose of this exercise is to answer questions for each strategy such as, "By how much will this strategy reduce our fixed costs?" "By how much will it reduce our variable costs?" and "How will it affect our sales drive and new-product development plans?"

Once we have completed the above checks and have obtained approval for kicking off all the planned projects, actual improvement activities based on the system depicted in the TP deployment diagram begin. As shown by arrow 4 in the diagram, we formulate plans for implementing the strategies and then proceed to carry out and follow up our plans as shown by arrow 5. The daily individual results produced by the activities are accumulated at the bottom of the diagram to correspond with the individual objectives noted across the top. This allows us to see our progress towards achieving the overall objectives for quality improvement, manufacturing lead-time reduction, cost reduction, and so on as the activities proceed, and enables us to assess the effectiveness of our implementation plans.

As a result of arranging the TP deployment diagram in this way, we obtain a clear idea of the strategies we must implement, these strategies having been worked out by taking our overall objectives, breaking them down into individual objectives, checking them in terms of customer satisfaction, and incorporating the ideas and opinions of the company's employees. All that remains after this has been done is to integrate the implementation of these strategies with our routine daily activities. The diagram also enables us to monitor our progress and keep a close check on how far the implementation of the strategies has moved us towards achieving our overall objectives.

Relationship Between the Bottom-Up Approach Employed in TP Management and the Go-Go Tools

Figure 6-2 is a simple outline diagram with product power and market demands set as overarching TP themes. Quality improvement, delivery-time reduction, and cost reduction are established as the overall TP objectives, and these are deployed by cascading them down to each individual process and working out their numerical expectations and contributions. Ideas for achieving the individual objectives are collected from the people in the workplace (that is, from the bottom up), and the pocket matrix technique is used to involve everyone in this process.

TP management is not a rigid system, and each TP prizewinner has a unique approach to its implementation. However, all these companies are extremely well-organized in the area of collecting ideas from their employees, and their overall small-group activity programs all run extremely efficiently.

It is my sincere wish that all companies will incorporate the TP management approach in order to strengthen their small-group activities. The purpose of small-group activities is to develop people's capabilities and encourage them to pursue ever more challenging targets. I hope that ways will be found to express this important people component visually and incorporate it into the TP deployment diagram.

BOOSTING THE PERFORMANCE OF SMALL-GROUP ACTIVITIES THROUGH FIVE-STEP EVALUATION (PTESC)

One approach used in Japan for establishing small-group activity themes and upgrading the level of the activities is PTESC (planning, trust, example, selection, and creativity). The key points in the PTESC approach are to plan effectively, win the trust of subordinates, set a good example, select the best ideas, and exercise creativity. Another popular approach used by improvement facilitators in Japan is the ORIENTT approach (observe, record, invent, enumerate, test, and try). In this approach, we observe and note down what is presently happening, devise ideas to improve the situation, calculate the benefits that will be achieved by implementing the ideas, choose the best ones, carry out any tests necessary, and then try to put the selected ideas into action.

The PTESC approach is very important for effective small-group management. Managers start by planning their strategy, that is, establishing objectives and deciding how to proceed. This is of course the first step to take in carrying out any project. The second basic requirement is to secure the trust of one's subordinates. Managers should tackle their improvement projects while doing their best to improve their own characters and personalities and win the confidence

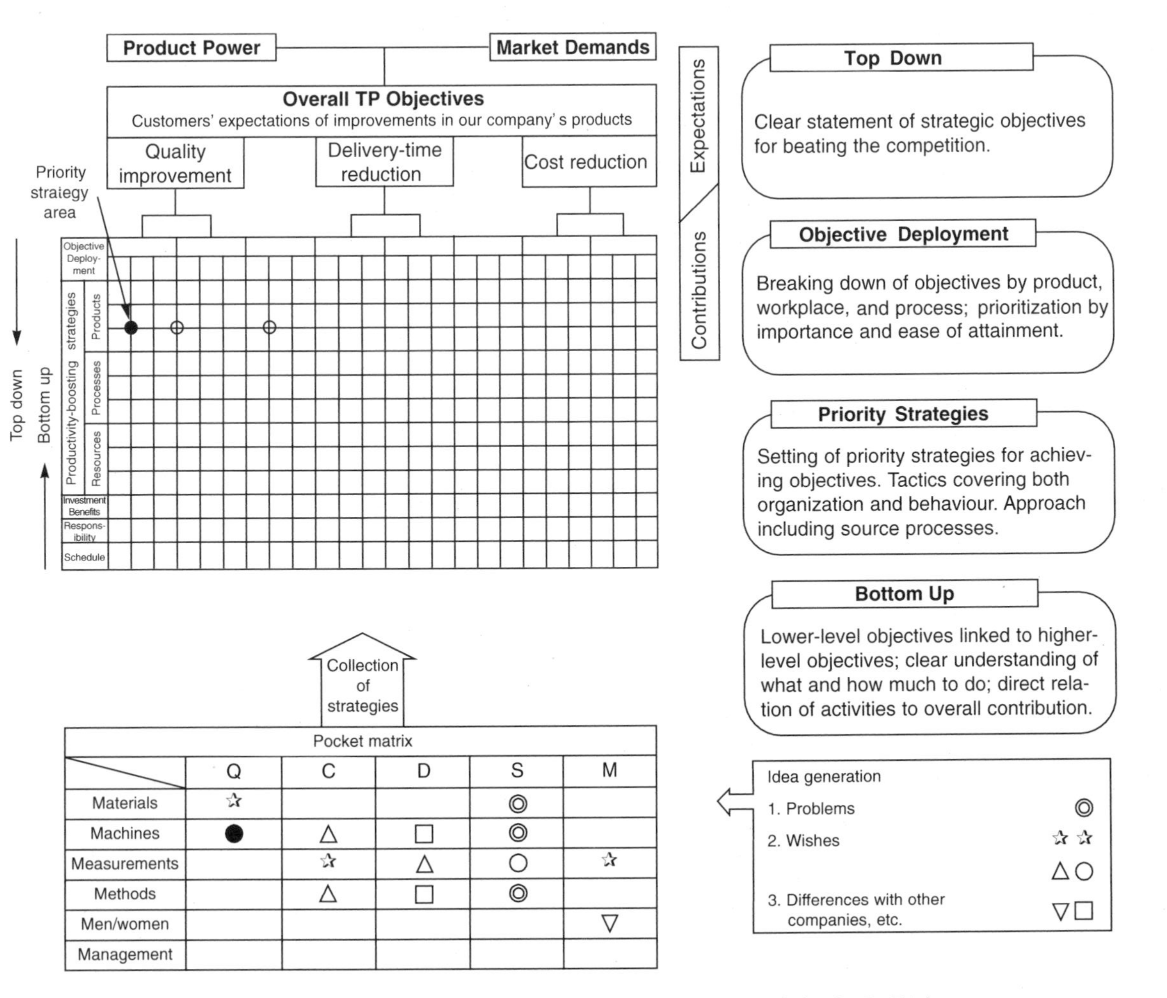

Figure 6-2. Relationship Between the Bottom-up Approach in TP Management and the Go-Go Tools

of the people under them. The third basic requirement is for managers to take the lead in solving problems. When people see their leaders (especially top managers) taking the initiative, they tend to feel that they must follow their example. Senior managers must demonstrate by personal precept the direction they want their company to move in, since this is really the only method by which they can get everyone moving in the desired direction. The fourth basic requirement is to select the best plans from all the ideas available. It is the duty of managers to pick the best means of achieving the objectives from all the good ideas presented to them. The fifth and final step is to exercise creativity and ingenuity. This means being original and inventive in coming up with effective topics to pursue and working out what needs to be done in order to improve the situation.

Six Additional Requirements for Good Management

There are six requirements for good management (often cited in management textbooks) in addition to the above five key points.

1. Set challenging targets. However if the targets are too difficult, they will take too long to attain. In such a case, set a large number of intermediate targets by dividing the schedule up into a master schedule, intermediate schedules, and daily schedules, and monitor and follow up progress on a daily, weekly, and monthly basis.
2. When targets are not achieved, do not merely write out death certificates or post-mortem reports.
   - Take prompt action whenever there are signs of danger.
   - Sound the alarm whenever things are getting out of hand.
   - When a target is not achieved, immediately investigate and plan countermeasures. (It may occasionally be necessary to change the plan; however, make every effort not to alter the master schedule.)
3. Treat the implementation stage (that is, the execution of work in a manufacturing operation) as a chance to discover improvement opportunities. When implementing a plan, discover sources of time-wasting while you work. As you proceed, reflect on time-consuming and difficult tasks and treat them as elements of the job that need to be improved. Then incorporate the results of these deliberations into the next round of tasks. Try to keep on improving your work, just as you would keep on trying to improve your performance if you were playing a sport.
   - In addition to reflecting on what went wrong when more time was taken to achieve the target that had been planned, it is also important

to reflect on why things went well when everything went according to plan. List the reasons for this, sort them out, and use them in the next round of activites.

4. Including workers (and, depending on the task, people from suppliers or affiliated companies) in the planning process makes it possible to discuss things from a variety of viewpoints, since different people have different experiences, encounter different practical problems, and look at things in different ways. It will also give them a feeling of participation and ownership of the plan. Besides making them more aware of what is going on and improving their understanding and motivation, this also helps things go more smoothly when trouble occurs or when minor adjustments are required at the implementation stage.
5. It is important to define the system of responsiblity and delegate authority in the plans. Ensure that the responsibilities for each job and each leader are clearly assigned and make sure that, as overall leaders, managers pay attention to only the most important areas. Each subleader should consult the overall leader only when an abnormality or emergency occurs.
6. Do not forget to use the principles of visual management to create an organizational setup that stimulates competition and allows people to experience a sense of achievement. Ensure that suitable events are planned to mark progress. It is important to divide the work up into definite stages and celebrate the completion of those stages.

We will discuss the significance of each of these six requirements.

Requirement one: Set challenging targets

As mentioned, if these are too high, they will take too long to attain. It is important to break the overall target down into detailed intermediate targets and keep a constant monthly, weekly, and daily check on progress. In this way, reliable progress will be made towards the achievement of the final target, one step at a time. This system, sometimes known as "milestone planning," is extremely effective when running a program of small-group activities. Once the team has reached a particular milestone, it sets sights on the next, and so on. The idea is to ensure that people experience the satisfaction of achieving definite goals as they move towards the attainment of their final target.

Requirement two: Do not abandon a target if it is not achieved

We want to keep our targets alive, not write out death certificates for them. It is important to take immediate action whenever there is any indication of a problem in the offing. This means that we need to set up an alarm system to warn of potential trouble. Managers should keep a close watch on the activities of their small groups and be prepared to step in and give assistance whenever the groups encounter difficulties; and the sooner this is done, the better. However, if conditions change and a small group is unable to proceed with its project, it may be necessary to change the topic, the group members, or the way in which the activities are being conducted.

Requirement three: Look for improvement opportunities while implementing the plan

The significance of this is as follows: Small groups are naturally keen to complete their projects and achieve their goals, and this is of course important. However, it is equally important to do more than just pat ourselves on the back once we have completed an improvement project. In a sport such as football, the players improve their performance in a game by observing what is happening and by changing their plan if it is not working. It is the same in manufacturing or improvement activities; people should be constantly on the lookout for problems as they go about their work. They should think about what they are doing and make a note of any problems so that they can use these as improvement topics in the future. Athletes constantly monitor their own performance and practice to eliminate their weak points. Workers on factory floors or in offices should do the same. We want to create a cycle of improvement followed by practice followed by further improvement, and the faster we can spin this cycle, the better. By speeding up the plan-do-check improvement cycle, not only do people experience the joy of success more often, they also sharpen their problem awareness and strengthen their improvement abilities.

Requirement four: Include the people who are actually responsible for carrying out the work in the planning process

I have repeatedly emphasised this in this book. Managers should make every effort to involve the workers at the planning stage in order to benefit from a wide variety of viewpoints and give the workers a better appreciation of what is going on. An understanding of the need for a particular action (the "What?" and the "Why?") is extremely effective in speeding up the solution of problems.

Requirement five: Managers should define people's responsibilities when plans are formulated

From the manager's viewpoint, this means delegating authority appropriately. Asking group leaders to do certain things is the same as dividing up the work and delegating authority for the various tasks.

Requirement six: Managers should set up a system that clearly shows how well the task is being accomplished

This is one of the principal purposes of the TP deployment diagram, which shows the details of the activities being undertaken and the results achieved, thereby making it easy to monitor progress and evaluate performance. Managers should also plan special events, such as presentation meetings, to mark their small groups' achievements. However, a presentation meeting should not be an end in itself; it should be regarded as a means of rapidly communicating small groups' achievements to the rest of the workforce. It is important to hold such meetings with the emphasis on letting people appreciate the satisfaction gained from the successful completion of an improvement project and to help them understand the nature of the improvements achieved. There are probably many other ways of achieving the same objectives, such as making public announcements, holding study tours to view the improvements, publishing the details on wall newspapers and so on. The method is irrelevant as long as the objectives are achieved.

The six requirements described in the previous paragraphs are important guidelines that all managers should follow to run a program of small-group activities.

USING A FIVE-STEP SCALE TO EVALUATE SMALL-GROUP ACTIVITIES

The next part of this chapter introduces my five-step scale for evaluating small-group activities, as shown in Table 6-1 on pages 78-79. How should managers go about setting challenging targets and raising the level of their small-group activities? I believe this five-step evaluation system can be extremely effective in achieving this.

As the table shows, the condition of a company's management has been split up into a number of elements, such as the 5Ss, workplace discipline, management by objectives, and so on. Each of these is evaluated on a five-step scale ranging from level 1 (the lowest level; for example, producing low-grade goods by a labor-intensive process) to level 5 (the highest level; for example, world-class manufacturing). The table contains a description of what constitutes each of the five levels for each of the elements.

The verbal description of the level for each element given in the five-step evaluation table could be replaced by a cartoon, or by numerical data. The elements could also be further subdivided—for example, the 5Ss could be split up into their individual constituents (sorting out, arranging efficiently, checking-through-cleaning, setting standards, and self-discipline). The point of the exercise is to decide on the specific items to be evaluated and use the data as a checklist to pinpoint one's present strengths and weaknesses and determine what action needs to be taken. The form in which this is done does not really matter provided that progressively higher goals are set in stepwise fashion. For example, if an athlete were using this method to assess his strengths and weaknesses, he might find that he had strong legs but relatively weak stomach muscles. He might then decide that he needed to work on strengthening his stomach muscles while doing everything possible to capitalize on the strength of his legs. Self-knowledge is an important tool. A radar chart is also sometimes used for plotting strengths and weaknesses, but this too is a tool, not an end in itself. It is only a way of highlighting the problems and making them easy to see. The content of the chart or table used is more important than its form, and it is essential to use the evaluation process to draw up a menu of action to be taken.

The five-step evaluation process is thus used to clarify the topics that need to be addressed. It is used to find out the company's present level, let all employees know the areas in which improvements must be made, and motivate people to make every effort to better themselves and their work. It is also important to keep improving the evaluation process at the same time as devising ways of improving employees' capabilities. The preparation of this kind of data is an essential precursor to any move to upgrade small-group activities.

Table 6-1. Five-Step Scale for Evaluating Small-Group Activities

Element	Level 1	Level 2	Level 3	Level 4	Level 5
5Ss	Unnecessary or spoiled items have been sorted out and disposed of.	Everything is in its proper place, and equipment panels and other surfaces have been cleaned.	The 5Ss have also been applied to the interiors of cupboards and storage areas with economy of motion in mind.	Materials and parts are ordered automatically, and sources of untidiness and contamination have been eliminated.	Materials and tools are supplied on a just-in-time basis, and management problems are made visible.
Workplace discipline	People manage their individual work well, regular morning meetings take place, and there is good all-round discipline.	Everyone assembles before the siren sounds, and people clear up properly after finishing their work.	People are dressed and ready to go and machines are up and running at the start of the morning shift.	Standard times are adhered to, the 5Ss are assiduously applied, and the whole production operation is carried out in a highly professional manner.	All goods are produced right the first time, and all processes start up smoothly in synchronization with the sound of the morning siren.
Management by objectives	Macro-type management by objectives is practiced, with the actual management of the work left up to the workplace.	Individual workers produce to daily objectives based on standard work times.	Standard times are strictly observed, in conjunction with activities to eliminate waste.	Clear overall and individual objectives have been set, and the support system is so flexible that people from one workplace can be sent to help in another workplace on an hourly basis.	Clear top-management and individual objectives have been set, problems are tracked down in the workplace, and workplace improvements are proceeding smoothly.
Using multiskilled workers	Each process has its own specialized workers, and very few are able to work on other processes.	Observation and waiting are treated as waste, and efforts are made to use this time for useful work.	Charts showing multiskill levels have been prepared, skills training is provided, and positive efforts are being made to increase the number of multiskilled workers.	Efforts are being made to learn new technology in order to enable each worker to handle three to five processes.	Every worker can freely handle five to six processes, and workers rapidly become experts at their new jobs when transferred to other workplaces.
Small-group activities	The management of small-group activities depends on the skills of individuals, and no activities are undertaken outside regular duties.	Everyone understands the significance of small-group activities, and there is a proper system for managing them.	Group members take turns as leaders, and improvement techniques are used for improving quality and work practices.	Exchanges take place with groups from different workplaces, and improvement campaigns are producing good results.	Small-group objectives are integrated with factory objectives, and new issues are constantly being addressed.

Element	Level 1	Level 2	Level 3	Level 4	Level 5
Process control	Workplace production management consists merely of noting priorities.	Production orders are issued and tracked for each process based on the use of standard times.	Daily production schedules are strictly observed, and level production is being pursued.	Small-lot production is being achieved through the use of the just-in-time system.	Automation of control by means of FA and FMS systems is being pursued.
Improvement-team activities	The improvement promotion system consists only of submitting suggestions and waiting for them to be implemented.	Employees make simple improvements on their own initiative beyond working hours.	Improvement teams exist, and dedicated groups are engaged in acquiring skills and implementing improvement suggestions.	Improvement teams are actively acquiring expertise in unattended automation and low-cost automation.	Improvement teams are capable of building equipment that makes use of the distinctive characteristics of the workplace (for example, constructing robots).
Identifying improvement topics	Employees are receiving training in spotting wasteful, non-value-adding work, and are beginning to make suggestions as to how to eliminate it.	Techniques for improving work practices, quality, and equipment are being applied in the workplace, and waste is being discovered.	Improvement techniques are being applied in combination (IE + QC, QC + TPM, etc.; for example, utilizing QC and safety techniques in TPM).	Groups are freely able to apply techniques such as error-proofing for preventing problems at their source (the principles of the three actuals is applied, and improvements are made in the workplace).	Groups are studying production engineering itself, and improvements and workplace knowhow are being extended to collaboration with engineering staff to improve the work done at the design and development stages.
Affiliate company training	Affiliate company training is aimed exclusively at preventing defects and other problems.	Teams are sent to affiliate companies to give specific advice, and bonus and penalty systems are skillfully applied.	Affiliate companies are improving their capabilities by organizing study groups, plant tours, and technical exchange meetings among themselves.	Affiliate companies are setting objectives and developing activities in the areas of the 5Ss, JIT, and so on, in collaboration with their parent company.	Affiliate companies have systems that allow them to stand on their own feet and they are acting independently.
Quality assurance and TPM	People are classified only as operators or inspectors, and countermeasures against defects are implemented only about once a month.	Activities to build in quality via the process have begun, using self-management, mutual inspection, patrols, and so on.	The seven QC tools are in use, and antidefect measures such as error-proofing are being implemented.	Activities designed to build in quality by means of the equipment are taking place, and efforts are simultaneously being made to eliminate sources of defects and improve the equipment.	Improvements are being introduced at the source of the production process, and everybody is studying and implementing zero defects and zero breakdown systems.

CHAPTER 7

Psychological and Scientific Principles of Small-Group Activities

I should now like to discuss why small-group activities came into being, what psychological and scientific principles they utilize, and the principles on which any drive to upgrade people's individual capabilities should be based.

Let us first examine the principles by which a small group comes into existence. Figure 7-1 illustrates how a small group begins, and how human qualities are exercised by setting up a small-group system. An experiment was conducted in the United States on a company of about 200 employees working on the production of documents. The problem was that the defect rate was extremely high at this company. All the employees worked together in one large room and were referred to by number rather than by name. Of course, they were not divided into groups.

One day, when the defect rate had risen to an unbearably high level, the president of this company sought the advice of a certain university professor. In the past, this professor had studied the career of the famous Japanese general Toyotomi Hideyoshi and had researched the story of how this general had had the walls of a certain castle built in three days. The professor had studied the principles by which a sense of responsibility can be engendered and results achieved in a remarkably short time when a workforce is divided into small groups. He had concluded that a system of responsibility develops when work is shared. On being asked to advise the company, the professor decided to try applying the principle of small groups to its operations. This experiment is said

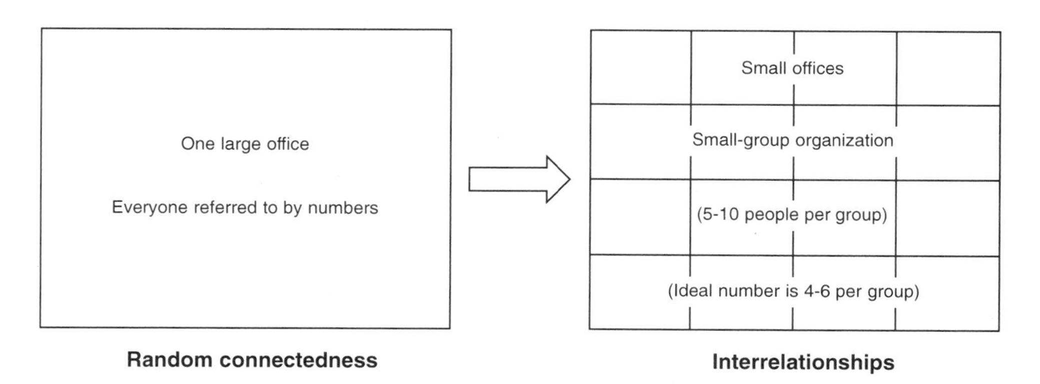

Figure 7-1. How a Small-Group System Brings Out Human Qualities

to have helped in building the set of principles on which today's small-group activities are based.

So, what specifically took place? The professor divided up the 200 employees of the company into groups of five to ten people. He abolished the practice of calling employees by numbers and made sure that everybody was addressed by name. He appointed a leader to each group and simply told the groups' members to help each other while doing the job as well as they could. As a result, the defect rate fell dramatically.

In the process of this experiment, many techniques for running small-group activities fell into place. In particular, clarifying each individual's role was effective in generating motivation, a desire to improve, and a sense of responsibility. It was also reported that the system of having employees in each group help each other while competing with other groups provided excellent stimulation and led to improved work results. This was the outcome of the American professor's scientific analysis of Toyotomi Hideyoshi's management methods. This research continued, and companies in Japan also began to make use of small groups, producing notable results. All this took place over twenty years ago.

MASLOW'S FIVE-STAGE HIERARCHY OF HUMAN NEEDS

I will now talk about Maslow's five-stage hierarchy of human needs (see Figure 7-2) in the context of linking individual needs to social needs.

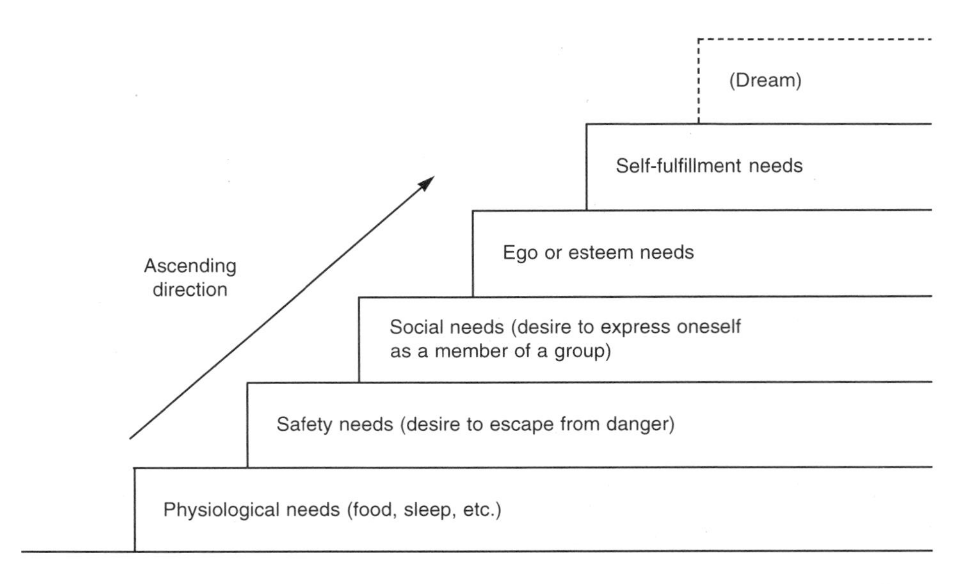

Figure 7-2. Maslow's Five-Stage Hierarchy of Human Needs

To begin, human beings cannot do anything unless their physiological needs are satisfied (stage 1). We cannot live without food or sleep, and it is therefore essential to meet these needs. People will face great danger and hardship in the attempt to satisfy them.

Once our physiological needs have been satisfied (we may keep our own animals for food or sow seeds and cultivate crops, for example), we no longer risk life and limb to go chasing after something to eat. Instead, we look after our own security. This is stage 2 of Maslow's hierarchy.

Once we have satisfied our stage 2 needs, we begin to think about satisfying our social needs; that is, taking part in communal life (stage 3). We desire to work with other members of a group, as illustrated by Figure 7-1. We begin to want to belong to a group and fulfill a role within that group. We desire to operate in a situation where we are accepted by others and are treated as individuals (rather than as faceless numbers). In a company, forming teams at this stage of the hierarchy permits people to exercise their individual talents and leads to the satisfaction of both individual and corporate needs.

We now move on to stage 4. We want more than just the sense of security we feel from belonging to a group; we also want to be respected by the people around us. This is one of the principles utilized in running small groups. When a topic is decided, results are reported at presentation meetings, and the group members are praised for their efforts.

The 5th and final stage of Maslow's hierarchy is the stage at which the need for self-fulfillment emerges. This implies setting our own objectives and proceeding towards their attainment. It is what happens when a person studies extensively on his or her own for the best way to perform a particular job. It resembles the process of becoming a professional in a particular vocation. People at this stage will attempt to satisfy their needs by working on their own to achieve self-set objectives, even when what they are doing is not what their company is engaged in.

This is Maslow's classical five-stage hierarchy of human needs. However, many people now say that something else needs to be added to it. What is it that drives us to move up the hierarchy? Many people say that it is something we might call a "dream." In running a business, what we need to do is to create dreams that marry our own ideals to the goals of our companies. The idea of creating "dream factories" is a fashionable topic in industrial circles these days, and this is basically what I am talking about here. Many companies are currently studying this issue, without having obtained any clear answers yet. It is a topic that needs further investigation; however, I have introduced Maslow's five-stage hierarchy as one of the fundamentals of small groups because it is basic knowledge required for understanding their operation.

THE DRIVING FORCE THAT DEVELOPS PEOPLE

I have just mentioned the importance of creating a dream in conjunction with applying the fundamentals of small-group management. Figure 7-3 indicates the forces that drive human beings. According to this model, there can be no philosophy (guiding principles, values, or beliefs) without a dream, no vision without a philosophy, no ideal without a vision, no action without an ideal, no results without action, and no pleasure and satisfaction without results. It is the pleasure and satisfaction of achieving results that keeps motivating us to act, and, ultimately, no meaningful action takes place unless a dream has first been established.

This teaches us that having a dream is of vital importance, since it is the driving force which motivates us to action. I hope that all companies will weave a dream into their corporate cultures to make their small-group activities even more effective in the future.

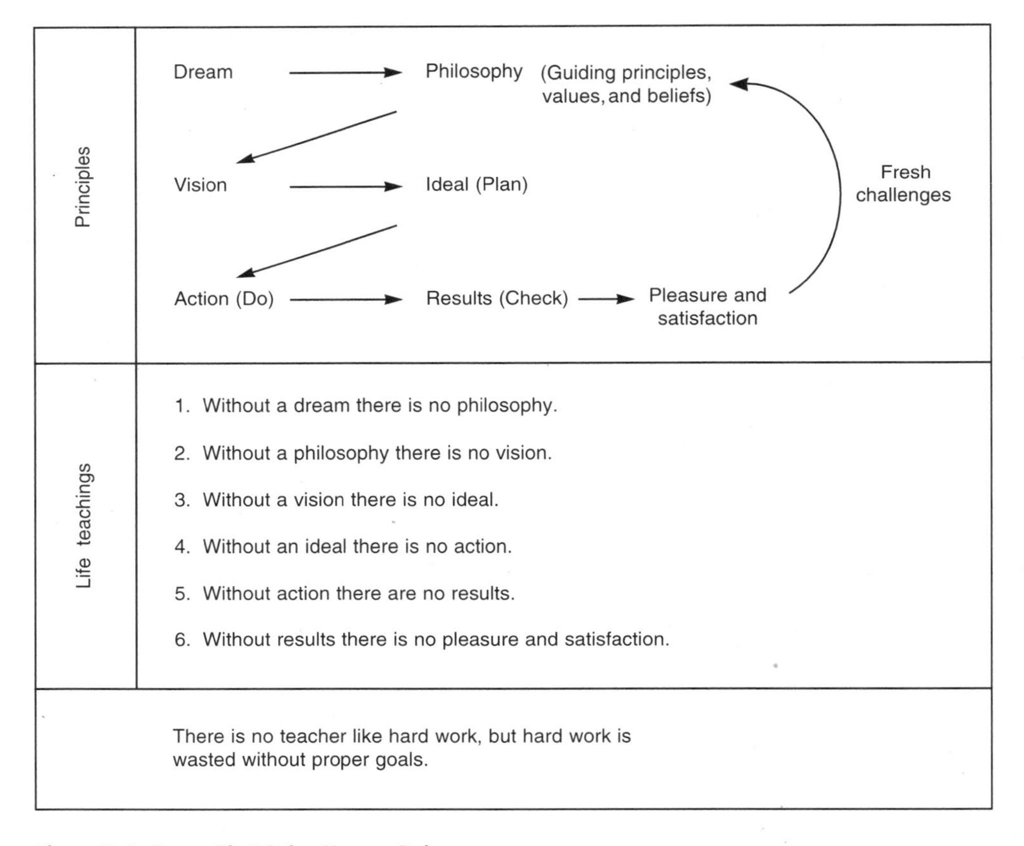

Figure 7-3. Forces That Drive Human Beings

MOTIVATION AS SEEN IN THE HAWTHORNE EXPERIMENTS

Figure 7-4 illustrates the results obtained in the experiments conducted by behavioral scientists at Western Electric's Hawthorne plant in Chicago, starting in the late 1920s. These experiments are said to have proved the existence of human motivation. The graph shows the relationship between productivity and the brightness of the lighting provided in the factory where the experiments were conducted.

When the scientists first went into the factory to conduct the experiments, they held detailed discussions with the employees and found that they were dissatisfied because their workplaces were too dark. The scientists therefore gradually increased the brightness of the lights. The productivity of the workers went up as this was done. However, just when the employees had begun to feel pleased with the company for improving the working environment to their satisfaction, the scientists began to dim the lights. Having first made them brighter, they began to reduce their brightness to the original levels. However, despite this, the workers' productivity continued to rise. There was no linear relationship between the brightness of the lights and the workers' productivity. Observing this, the scientists concluded that physical conditions were not the only factors affecting productivity, and that motivation plays a large part.

As the brightness of the lights was decreased, productivity began to decline only when it had become almost too dark for people to see what they were doing. The scientists concluded that people's productivity depends, to a large extent, on how much they feel valued by the organization they are working for.

In another experiment, the plant's employees were asked what bothered them most about the company's cafeteria. They replied that the food was poor, and

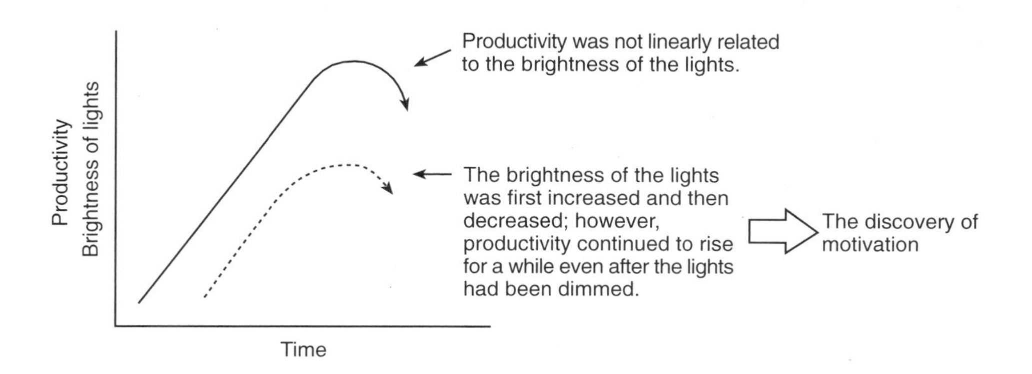

Figure 7-4. Proving the Existence of Human Motivation

they deserved better because they worked so hard. The scientists told the employees that they would approach the company's managers and ask them to improve the food and provide a more varied menu. In fact, however, they did nothing. They merely pretended to have done so. They then asked the employees how their food was on a particular day, telling them that they had asked the managers to make some improvements and that the day's menu was the result. They described specifically what was on the menu and how it had been cooked. The outcome of this subterfuge was that the employees asserted that the food had improved, and their productivity also went up.

The scientists claimed that this experiment showed that having one's requests accepted by the organization one works for is a major factor in improving productivity. Small groups often become motivated and energetic after top management has discussed corporate issues with them and involved them in the selection of improvement topics. This is an example of the use of the principle discovered in the Hawthorne experiments. Management is based on definite laws and principles, and these experiments illustrate how important it is to act on the basis of a knowledge of these laws and principles when managing small groups.

STRENGTHENING TEAMWORK

Figure 7-5 illustrates one of the well-known principles of management by which techniques used to improve sports results are used to boost teamwork within a company.

The left hand diamond of Figure 7-5 takes the game of cricket as an example. Winning teams can be classified into four types: teams that are very good at batting, teams that are very good at bowling, teams that are very good at fielding, and teams that display excellent teamwork. The top team in a league will be the one that consistently achieves the right overall balance among these four factors. If a team has a weakness in a certain area, as shown by the dashed lines, it will lose. It is said to be possible to pick the winning team in a competition by assessing each of these four characteristics on a five-point scale, plotting the values on radar charts, and comparing the charts.

Enhancing the activities of small groups is a key issue for companies, since it affects employee motivation, teamwork, and improvement capabilities. A company must achieve a good balance between enhancing small group activities and its product development capability, as well as its quality and delivery performance. It must also keep its costs down (in the cricket analogy, doing all this would be equivalent to fielding well). Enhancing the activities of small groups (the left side of the right-hand diamond) imparts vitality to the three elements at the other corners of the diamond. The achievements of a company are represented by the

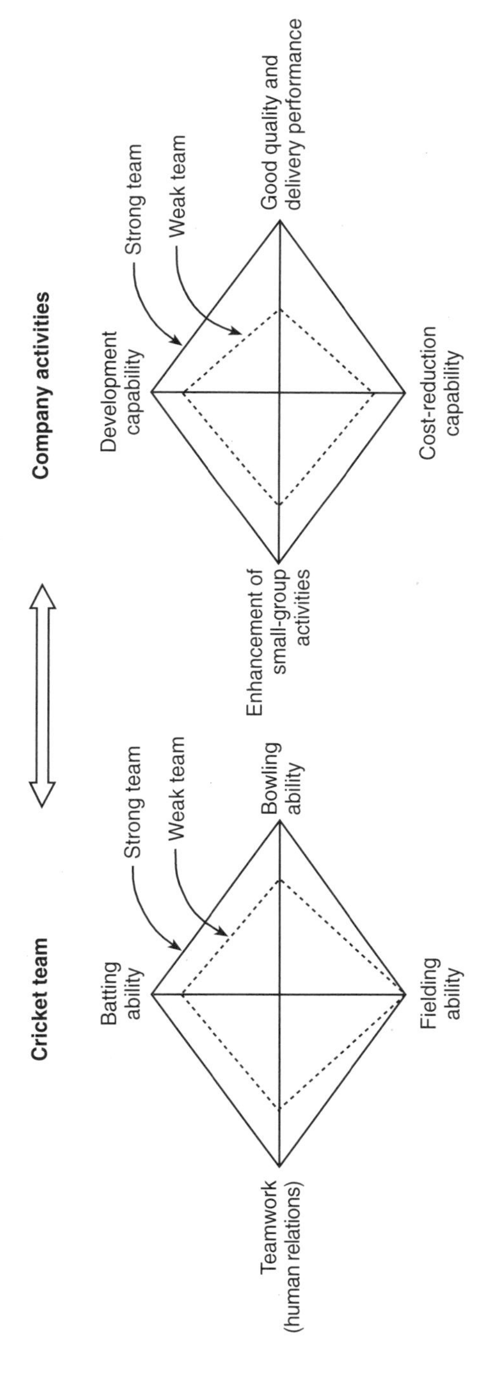

30 minutes of motivated team-building is better than 2 hours of enforced practice.

Figure 7-5. Applying Sports Management Principles to Improve Teamwork

dashed-line diamond shape moving to the right, powered by the small-group activities at the left. It is obvious from this analysis that teamwork is an extremely important element of corporate success.

A certain sports team manager says that he gets better results from thirty minutes of practice when his players are motivated and working as a team than from two hours of practice when they feel compelled yet reluctant to perform. This really shows the importance of small groups. Practice makes perfect, and small-group activities seem to be the real driving-force behind a company's achievements, especially in the areas of improving development capabilities, boosting quality and delivery performance, and reducing costs.

MCGREGOR'S THEORY X AND THEORY Y

Table 7-1 outlines two sets of beliefs and assumptions about workers that McGregor entitled Theory X and Theory Y. In the previous chapter, I suggested that the best way for managers to develop the abilities and confidence of the people under them is to praise them, not criticize them. This is the approach advocated by Theory Y. In contrast, Theory X assumes that workers are inherently lazy. Consequently, they will try to shirk their work unless they are compelled to do it in various ways. In the Theory X approach, workers are forced to stay at their places of work and do as they are told.

Basic Assumptions of Theory X	Basic Assumptions of Theory Y
1. Most people have an inherent distaste for work and try to avoid it if at all possible. 2. Since most people are born with a dislike of work, they will only exert themselves sufficiently to achieve their company's objectives if they are coerced, controlled, ordered, and threatened with punishment. 3. Most people like to be given orders and tend to shirk responsibility. They have little ambition and seek security above everything else.	1. People are born with a desire to devote themselves to their work in the same way as they do to sports and leisure pursuits. 2. Control and intimidation are not the right way to get people to work towards the achievement of corporate objectives. People will drive themselves to achieve their own self-set targets. 3. Whether or not people devote themselves to the achievement of their objectives depends on the rewards they receive for achieving them.
(Work approach) Compel people to work by authority, brute force, intimidation, and coercion.	(Work approach) Respect people's inherent needs for self-esteem, a sense of responsibility, and the desire to improve.

Table 7-1. McGregor's Theory X and Theory Y

Theory Y, however, adopts the standpoint that human beings are inherently good. In this approach, it is believed that workers are well aware of the problems affecting their work and are willing to try to better their own performance. If given the opportunity, they will study by themselves and rapidly improve their work. This is obviously a very different philosophy from that of Theory X.

At one time, Japanese companies used to hire as many university graduates as they could and employ them to observe the workplace and improve it by using complicated academic statistical theories. However, Theory Y is said to have changed all this. The application of small groups in conjunction with Theory Y produced a major upheaval in the way companies operated, for the reasons explained in the following paragraphs.

Imagine that an engineer is working at a company. Although he is highly competent, he has many things to think about and can only apply about 70 percent of his time to improvement projects. He has many other things to do apart from thinking about improvement, such as writing reports, attending meetings, talking on the telephone,and so on. In practice, therefore, even if he is assigned full-time to improvement work, he can only spend about 70 percent of his time on it.

Now visualize a workplace in which 100 people are employed, and assume that these people spend only .01 percent of their time on improvement. This is a very low figure. However, let us imagine that a program of small-group training is started, and the employees request that they be allowed to spend .01 percent of their time on improving their workplace by themselves. Applying Theory Y, the engineer imparts his knowledge in an easily understandable form to the employees in the workplace. When this is done and the workers set about collecting data and tackling problems, the amount of output is not simply .01 percent x 100 = 1 whole unit of work, it is more like 2 or 3 units. This was *actually* experienced in practice, and it gradually became apparent that small groups could accomplish the same kinds of improvements that previously required a full-time engineer.

Although the highly paid engineer in this example was achieving an input of around 0.7 units (or 70 percent of his potential input), the hundred workers, each putting in .01 units of effort (for an input of 1 unit) were actually achieving an output of not 1 unit but of 2 or 3 units. In other words, the small-group activities were producing the same level of results that employing two or three full-time engineers would have. This is because three full time engineers at .7 percent apiece only produce 2.1 units of work. In fact, this research into Theory Y and small groups by Japanese companies revealed that the output kept increasing beyond this level as small groups became better and better at making improvements.

What this suggested was that sending engineers to the workplace with instructions to identify problems leading to defects is not likely to succeed because the kinds of problems large enough to show up in statistics do not happen all the time. Recent defect rates are of the ppm order, and there are very few opportunities to see problems happening. However, the people in the workplace are constantly in touch with the situation and see it from a variety of standpoints. Since they are in the workplace all the time, they are far more likely to see problems happening than engineers who only visit the workplace occasionally. If the people in the workplace are given the knowledge to identify and analyze problems skillfully, they can, in fact, be far more effective than engineers, since they have so many more opportunities to identify and effect possible improvements.

This is what gave rise to the notion of leaving the workplace up to the workers and letting the workplace manage itself, and these ideas were embodied in the F Plan (a system of plant management by foremen and forewomen) proposed about fifteen years ago by the JMA. In this system, people who have done their training in the workplace and their learning on the job, find and solve workplace problems by themselves, and workplace leaders develop the people under them through these improvement activities. The system was found to increase the ability of the people in the workplace to effect improvements. When the types of problems the people in the workplace were solving were analyzed, it was found that they consisted of problems these people had known about all along. They had merely been left unresolved because the people in the workplace had not been afforded the time or knowledge required to solve them.

It was therefore decided to form small groups of workers in order to tackle workplace problems, and this approach was continued for the next decade and more. The point is that workplace problems are usually best solved in the workplace. Also, since the problems are being solved by the people who are most familiar with them, the most logical and effective solutions are usually found. This was how Japanese-style small-group management, based on Theory Y, came into being. From this point of view, McGregor's Theory Y made a significant contribution to Japanese management practice. In particular, I feel that it would not be an exaggeration to say that the idea of developing the workplace's strengths played an extremely important role in improving management effectiveness.

THE 2-6-2 PRINCIPLE

Figure 7-6 encapsulates a discussion originally presented in a video available from the JMA. It illustrates a principle that applies to any group of people of more than about ten or twenty members. In fact, the larger the group, the more closely the principle has been found to apply. (Other group principles have also been identified; for example, there is the 7 percent principle, which says that in any organization, only 7 percent of the members act as leaders. These people constantly try to influence the others to go along with them, make them aware of the problems, and suggest issues for them to tackle. The nonleaders simply get on with addressing those issues.)

Returning to the original discussion, the 2-6-2 principle is an important concept that can help us understand the nature of groups and how to utilize them effectively in managing our organizations. Whenever something new is to be done, approximately 20 percent of the people in a group will try to find out how to respond and will take up the challenge on their own initiative. Meanwhile, about 60 percent of the group's members will watch the leading 20 percent, ready to follow them if things go well. The remaining 20 percent tend to put their energy into opposing the change. Managers responsible for handling small groups should be aware of this principle.

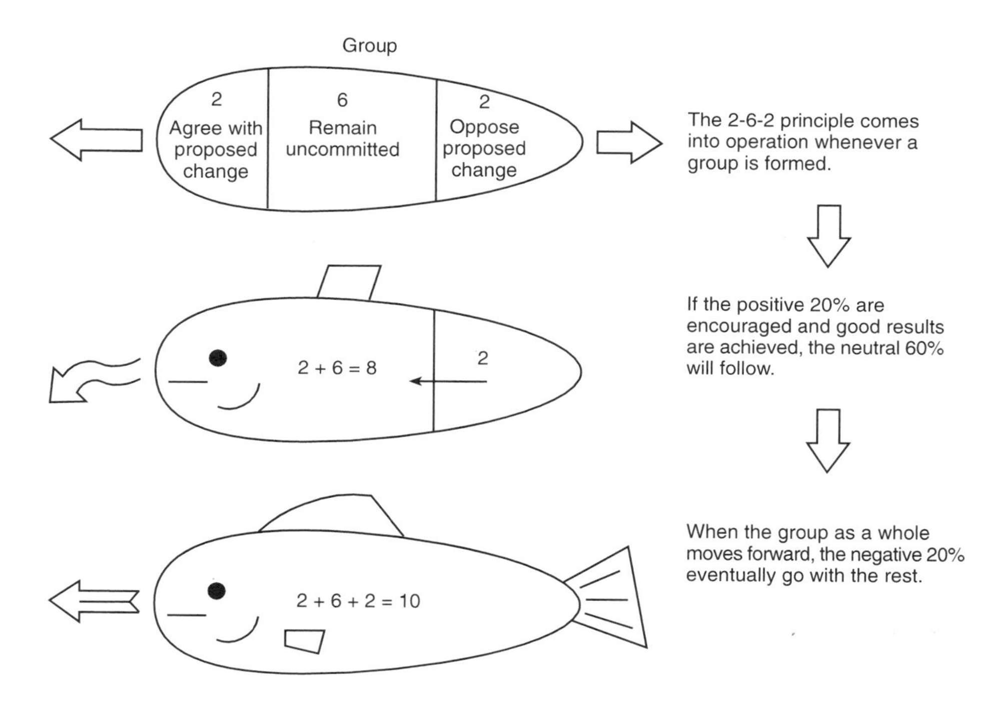

Figure 7-6. The 2-6-2 Principle of Group Behavior

To get a group moving, we should first be aware of the 2-6-2 principle. We should also realize that we will see little return on our effort if we concentrate on trying to convert the negative 20 percent to our cause. It is actually not a good idea to expend much effort on that 20 percent. It is much more effective to support the enthusiastic 20 percent by providing them with any training they need, helping them achieve good results, and recognizing their efforts. If this approach is adopted, the positive 20 percent are bound to perform well and produce results in a short period of time. The people who achieve good results are said to form the base of a company's training system. If the well-motivated 20 percent are praised, and the good things they have done are communicated to the rest, the people around them will take note of what is happening and try to emulate them. In other words, the 60 percent who originally adopted a wait-and-see attitude will notice what is going on and join in. Therefore 80 percent of the workforce will now be well-motivated and enthusiastic, with only 20 percent still dragging their feet.

In this way, the leading 20 percent get the uncommitted 60 percent off the fence and moving. If things go well, the negative 20 percent will then see the good results that have been achieved and will get going themselves and join the rest, without any particular training or persuasion. The 2-6-2 principle reminds us that whenever times change and progress is made or upheavals take place, groups always exhibit this characteristic. Managers responsible for small-group activities should be well aware of this principle when spurring their groups into action.

A manager of small groups always has the responsibility of keeping a number of teams active. The 2-6-2 principle does not necessarily apply to individuals, but it definitely applies to groups. As already emphasized, it is vital for small-group managers to recognize this principle and use it in leading their teams. Moreover, this way of handling small groups will definitely lead to improved company performance. In conjunction with the application of Theory Y, managers should motivate their small groups by praising the efforts of those who have achieved good results. This will create a driving force that will pull everybody along and will result in the development of people's individual capabilities.

The Application of the 2-6-2 Principle to Sports

The 2-6-2 principle is often used in training sports teams. Strong teams generally apply it at the same time that they fulfill a number of other conditions. I would like to give three examples of those conditions here.

Condition one: Every member of the team understands and fulfills his or her own role

All the team members strive to achieve their own objectives and do their best in their own individual capacities at the same time as helping one another. The players' roles are clearly defined, well understood, and faithfully performed.

Condition two: Team members are uniformly good at their chosen sport

They have a strong sense of professionalism, and the players' individual skills are highly developed. Although each team member is used in the most appropriate capacity, each could also perform satisfactorily even if required to play in a different position. When a team's members are virtually interchangeable like this, the team has achieved a high all-round standard. This can easily be seen in rugby teams, soccer teams, and cricket teams. If all the team members are highly skilled and they fulfill their appointed roles well, the team will definitely be a good one.

Condition three: Each player is given a particular aspect of the game to work on so that the kind of team just described can come together

If this is done, the team will be able to perform unexpectedly well in a crisis. In sports, 1 + 1 does not equal 2. The output is more like 3. Some teams are capable of producing something out of nothing; if the members trust each other and help each other, and if they know what they have to work on in order to improve their own abilities, they become capable of extraordinary achievements when it is demanded of them. This phenomenon also occurs in companies, when employees band together to tackle important improvement projects or secure major orders. Teams can also do great work when tackling problems such as delivery, quality, and so on.

Let us examine the first of these three conditions a little more closely. In sports, every day is a battle. Every run, goal, or point is important, and the team members concentrate hard because the score affects their own individual records as well as their team's overall results. This is why sports team managers study the principles of teamwork so assiduously and try to use them to make their teams stronger. In this sense, the management of a sports team is akin to small-group management, and the team's coaches study and apply similar principles. Among the team's strongest members there will certainly be some good coaches, and they will try hard to make their team stronger.

The same sort of thing applies to individual sports such as golf. Jack Nicklaus has a tremendous golfing record and is still active on the veterans' circuit. He too has an excellent coach. His coach analyzes his game and takes videos of his performances. If Nicklaus' score goes up, the coach analyzes the videos, points out what Nicklaus is doing wrong, and does everything he can to eliminate the weakness. In his role as pupil, Nicklaus does what he is told and follows his coach's advice. This is one way in which he keeps achieving low scores.

In the case of a company, this is the same as carrying out training and immediately applying it on the shop floor. It is important to adopt a training style that ensures that whatever is learned is immediately put into practice.

As this indicates, the best kind of training is one that marries theory with practice. The most effective type of training system is one in which the necessary training is received exactly when it is needed and is immediately applied, producing useful results. This is why small groups need good trainers. Conversely, if a small group is saddled with an ineffective trainer, this will have an extremely negative effect on its performance. The group will concentrate on "papermaking" (preparing beautiful charts and tables) and playing with statistics. They will find themselves in a situation where they are spending a lot of time on training but what they have learned is hard or even impossible to apply. I believe that it is very important for companies to establish the kind of training systems in which good coaches that are able to give practical guidance are taught to train small groups, and the groups' members listen to what they are told and apply it.

As far as the second condition is concerned, strong team members are developed more quickly if more of the players try their best to develop their own individual skills. The top players in the top teams may appear to play effortlessly, but they usually train extremely hard out of the media spotlight. The best players love their training and keep mastering new skills by persistent practice.

Self-study is also important in companies. Sometimes, people study so hard with the aim of achieving their individual objectives that they even forget that what they are doing is work. This can happen when people have clear objectives. When members brush up on their individual skills in this way, it helps to create a strong team. It is impossible to develop a team in which all the members are highly competent due to force. This phenomenon occurs only when the aims and objectives of the individual team members match precisely those of the organization they are working for. This illustrates the importance of promoting small-group activities to create a corporate culture and dream that employees can buy into.

To elaborate on the third condition, good teams make a point of testing their strength against teams from other leagues or geographical areas. Athletes in one discipline sometimes even pit themselves against athletes from other disciplines. They want to find out whether the effort they have put into honing their skills has paid off. They do this in order to confirm that they have attained their current objectives and use their achievement as a jumping-off point to seek fresh challenges. They always test the results of their efforts in competition. It is the same as learning something, practicing it by means of classroom exercises, and then immediately applying it in the workplace. It is important to keep repeating this cycle of learning, practice, and application.

Applying the Three Conditions to the Workplace

If we examine the three conditions required for developing skillful team members, we can draw the following two conclusions: First, people become good at what they like doing best; and, second, practice makes perfect. A workplace is not improved by the emergence of a single gifted person doing his or her best. It only really improves when all the people in the workplace try to improve because they want to do so for their own and their colleagues' benefit. This happens when they share a dream and do not begrudge the effort required to realize it. Workplace improvement is something achieved not through innate ability but through hard work, and this is the direction in which team members should be led. It is the duty of managers and team leaders to bring out the best in their people, and this is the whole point of small-group activities.

A program of small-group activities is an opportunity to develop people, and we gauge the results of our efforts in terms of how well those people operate their company's manufacturing processes and produce its products. This is where the similarity with taking part in a sports competition lies. The people doing the work train themselves and are helped along by their coaches. If these activities are taken seriously and a sense of unity and solidarity emerges within a corporation as a result, they go beyond being merely small-group activities and lead to the single-minded movement of the whole company in the right direction. In creating strong small groups and developing people, I believe it is essential to understand the laws and principles of small-group training well and apply them in the same way as in sports.

Chapter 8

Into the 21st Century with Small-Group Activities

THE HISTORY OF THE ZERO DEFECTS CAMPAIGN AND SMALL-GROUP ACTIVITIES IN JAPAN

The 30th anniversary of zero defects (ZD) activity was held in Japan in 1997 (see Appendix). The activity began at NEC Corporation in 1965 and developed at an unusually high speed. The number of sites engaged in the activity totaled 3,600 when the Japan Management Association organized the first national ZD convention in 1968.

ZD activity was first conceived in 1962 at Martin Corporation in Orlando, Florida. The whole company's manufacturing schedule needed to be cut drastically after receiving a request from the customer to shorten the delivery time by two weeks. It seemed impossible, but the company thought that by following working conditions and procedures correctly, from the beginning, they could do it. To make it possible, each employee had to be made fully aware of the importance of doing his or her job properly the first time. Independence and autonomy were necessary. The corporation implemented a campaign with the slogan "Do It Right the First Time" and achieved admirable success in shortening the delivery time.

This campaign to appeal directly to the minds of individuals was combined with traditional self-controlled small-group activities and developed into the present ZD activity and Japanese-style small-group activities. Small-group activities such as ZD activity became widespread not only in the manufacturing industry but also in other industries, thanks to the clear recognition of its value by people in industry. Furthermore, small-group activities spread from Japan and were introduced in foreign countries. However, Japan is now at a turning point in industrial structure and faces new problems and issues regarding the methodology of small-group activities. It has to be said that small-group activities have become stagnated to a certain extent. It is therefore necessary to review the course of the activity itself, while considering the future.

THE FUTURE DIRECTION OF DIVERSE SMALL-GROUP ACTIVITIES

The Changing Business Environment

Small-group activities were first started with the aim of stimulating workplace improvement activities. Backed by economic growth, they have spread across the industrial sector and have been changing greatly due to the following new trends in the business environment (see Figure 8-1).

Globilization

One major trend in Japan's industrial sector is progress towards globalization, such as shifting production sites overseas. At the same time, small-group activities are moving overseas, bringing up questions of how they should be organized in foreign countries with different customs and cultures. In Japan, too, more and more workplaces find that different nationalities are working together, requiring small-group activities to keep up with the change.

Increasing diversity of values

The second major trend is the diversification in people's values. Values affect attitudes toward one's own work, as well as toward the whole company, and many different values have to be respected. Small-group activities themselves

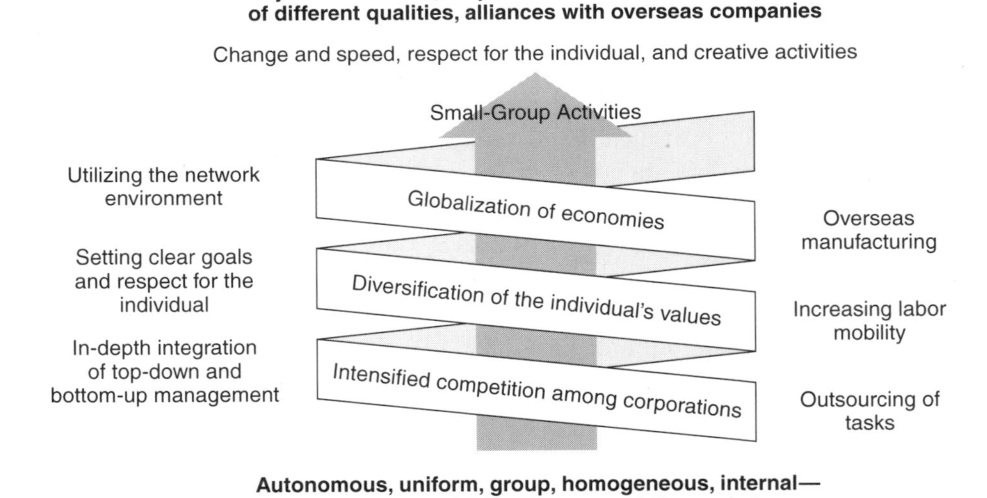

Figure 8-1. The Diversification of Small-Group Activities

need to adapt to this diversity of values. The traditional small-group activities tend to pay little attention to each member's strengths and weaknesses. It is important that small-group activities be implemented not by imposing conventional patterns, but by stating clear objectives with due respect to individual differences. It is not enough to merely continue the present form of activities.

Fierce corporate competition

A third trend is the intensifying competition among companies. The competition among companies in each industry is increasing. There is competition to establish de facto standards in high technology industry and the struggle to gain a larger share of mature markets. Under these circumstances, it can also be said that in order to win against the competition, small-group activities require the organizational power of the company as a whole. Therefore, small-group activities, which have been directed mainly from the bottom-up until now, need to incorporate top-down involvement.

Key words in small-group activities

Under these changing circumstances, small-group activities, which have thus far been characterized by additional key words such as uniform, group-work, homogeneous, and in-company, are being required to express themselves by additional key words and phrases such as objective-oriented, respect for the individual, integration of different qualities, and alliances with overseas companies.

New Formats for Small-Group Activities

Small-group activities in companies are already responding to the changing business environment by changes in the following activities (see Figure 8-2).

Changes in objectives

At first, small-group activities were started mainly to invigorate an organization by voluntary participation in improvement activities. Recently, however, small-group activities in some companies are being used to implement the changes that the participants themselves want to bring about. Sometimes the activities are directly aimed at improving division performance, and in some cases they are aimed at changing overall company management systems.

Changes in attitude or stance

Some companies have changed their attitude toward these activities from that of being satisfied with the participation itself, or by the fact that the activities are simply being carried out, to placing greater emphasis on the *results* of the activities.

Figure 8-2. New Formats for Small-Group Activities

Changes in themes

The range of themes has increasingly shifted from that of improvement within a team to that of improvement within the organization. Today's themes include improving work methods, finding ways to achieve division policies, and using themes dictated by company policies.

Changes in participating members

Just as the scope of activities has extended, the scope of membership has extended. It is often the situation that senior members of a team, who directly supervise it, also participate in activities. This participation often extends to before and after the activity and to associated departments, enabling activities with a wider perspective. Furthermore, some companies carry out activities involving related divisions, affiliated companies, and even client companies.

Changes in relationship to the organization or management

Small-group activities, originally established separately from the office organizational system, started with the principle that they were voluntary activities free from organizational limitations. Recently, however, there have been some cases where activities have been focusing on input from the organizational system, while maintaining the principle of voluntary participation.

Changes in relationship to other systems

In some cases, small-group activities are linked to other systems. For example, the themes of activities may be linked to results management. Small-group activities may be used as a means for improving the productivity of divisions. Therefore, staff participation and work in small-group activities is included in employee performance ratings.

Small-Group Activities Are Developing in Different Divisions

Small-group activities are not limited to certain sections of a company. They are growing and they are interconnecting (see Figure 8-3).

Expanding to all divisions

The kinds of divisions involved in small-group activities is diversifying. Small-group activities were originally used in production divisions and various techniques suited to the characteristics of those divisions were used. After significant results were achieved in the production divisions, small-group activities extended to all divisions in many companies. As a result, small-group activities now occur in divisions such as the engineering division, sales division, service division, and even in the personnel division. In nonproduction divisions, small-group activities must be adapted to that division's particular needs.

Alliances among divisions

Some companies have recently started tackling larger-scale issues by coordinating among different divisions, such as jointly selecting themes that are expected to maximize division results, cross-division participation, and so on.

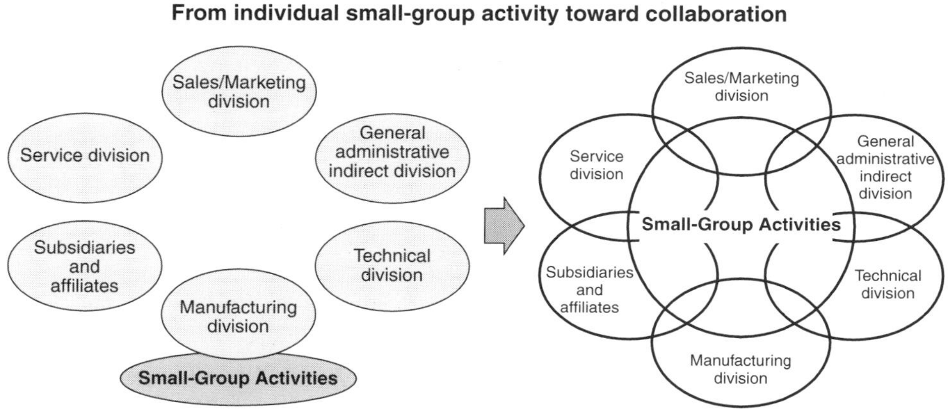

From individual small-group activity toward collaboration

Figure 8-3. Small-Group Activities Are Developing in Different Divisions

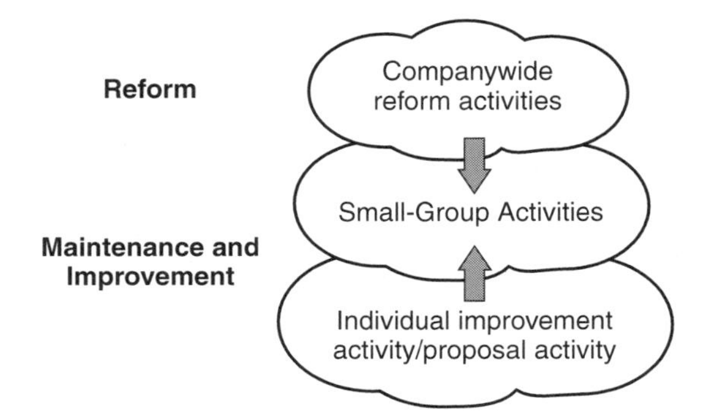

Figure 8-4. How to Position Small-Group Activities

How to Position Small-Group Activities

I would like to reconfirm the correct positioning of small-group activities. Small-group activities should be positioned between individual improvement and proposal activities and a company-wide improvement program that is carried out by a top-down process. When we think of the future development of small-group activities, we hope to take (1) the improvement and proposal activities performed by individuals and (2) the companywide improvements initiated by top-level management and combine them into small-group activities, while still maintaining the midway positioning of small-group activities (see Figure 8-4).

A CONCEPT FOR THE FUTURE DEVELOPMENT OF SMALL-GROUP ACTIVITIES

How should each company implement its own small-group activities? Should it use examples from other leading companies?

First we would like to note that activities are not static, but are constantly evolving. While we realize that one objective of small-group activity is to enhance its degree of contribution to corporate business, the diversity of small-group activities is attributable to three factors: the scope of activity, the scope for selecting themes, and voluntary participation (relationship to company personnel system), all designed to make a contribution to management. These factors are not fixed and are always changing with time.

A development map (see Figure 8-5) is a figure in which the just mentioned three elements are represented by assigning each factor to the X, Y, and Z axis respectively. The result of activities or the target (T) is shown as the product of the X, Y, and Z coordinates. It is understood that the further this value moves from its origin (O), the greater will be the result of the activities, and the greater the contribution to management.

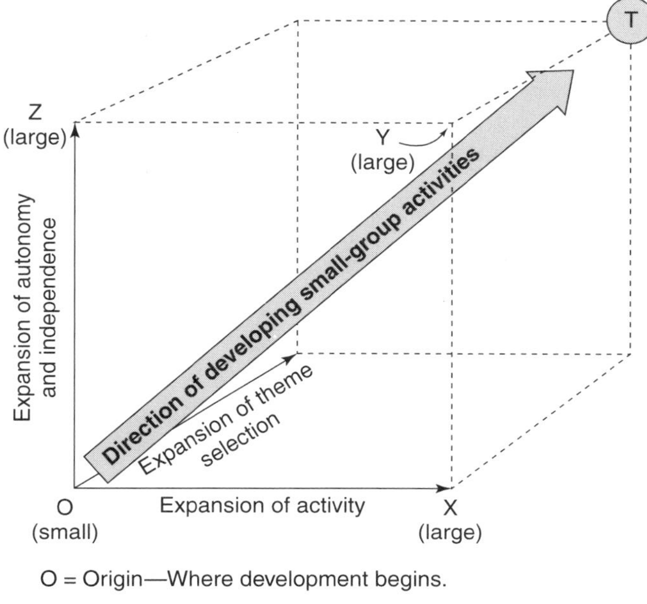

Figure 8-5. A Concept for the Development Stages of Small-Group Activities

When you begin to examine the future development of small-group activities, we would like to propose the following step. Position the current state of your company's activities on these axes and use those points to set objectives about which activities need improving and how much improving they need.

Steps for Small-Group Activity Development

As one guideline for positioning small-group activity development, we have prepared Table 8-1 to describe the development in four steps. Since this table shows only a basic pattern, it is not mandatory to use this when your company starts considering the positioning, but it can be used as a guideline. This table was designed by using the three coordinates of the development map just discussed.

Step one: The voluntary participation stage

This is defined as the stage right after small-group activities are introduced. It is where people participate in the activities and communicate with the other members but the emphasis is not on the way the activities are carried out or the

Table 8-1. Specific Practices for the Development Stages of Small-Group Activities

	Step 1 Voluntary participation stage	**Step 2** Self-actualization stage	**Step 3** Contribution to performance stage	**Step 4** Participation in management stage
General emphasis and characteristics of activities	• Uplift morale • Foster a sense of being part of the group—solidarity • Mutually understand the responsibility of others • Experience and exchange activities	• Reform individual awareness • Expand individual skills • Discover current problems and themes • Study improvement methods and technology	• Achieve task objectives • Integrate tasks with instruction given by manager • Expand scope for utilizing techniques and methods to promote improvements	• Continue to reform the task • Speed up the process • Establish projects • Complete the integration of top down and bottom up
***X axis:* Expansion of activity—work and scope (in relation to division)**	• Participate in activities with colleagues at the workplace	• Participate in activities with colleagues at the workplace • Superiors and staff members participate when deemed necessary	• Solve problems that go beyond the division • Collaborate with people involved in the same or similar jobs, neighboring jobs, and pre- and post-process jobs	• Get related divisions and affiliates involved
***Y axis:* Expansion of theme selection (in relation to policy)**	• Use independent themes • Improve things in the immediate surroundings	• Understand the division themes and share themes	• Use themes based on division policy	•Use themes based on company policy
***Z axis:* Expansion of autonomy and independence (in relation to management)**	• Receive guidance from the management	• Practice autonomous management activities	• Discover autonomous, independent themes and problem solving	• Promote self-reliance, autonomy, and independence, which is in line with the management

results. By promoting communication within the team, each member gets to know the jobs and problems of the other members and develops a working relationship with the other members, thus creating the sense of companionship and unity.

Relating this situation to the X axis, activities at this stage are mainly activities in the immediate workplace. On the Y axis, it is very likely that that the themes that will yield results most easily, or themes related to issues close to the members' workplace, are selected. As for the Z axis, the activities are limited to the specified task. At this point, if administrative or senior staff act against members' wishes, it is possible that the team will quickly lose interest in the activities. The key to continuing the activities is to make members feel that the activities are for their own benefit.

Step two: The self-actualization stage

The stage of voluntary participation is followed by the stage of self-actualization. In this stage, problems that arise within the team are fully grasped, suitable measures to solve the problems are developed, and improvements are made. The worries and problems of the individuals are resolved by this stage.

At this stage, the position of activities on the X axis is the same as in the stage of voluntary participation, but in some cases, senior staff or other staff members may join in the activities while improvements are made. As for the theme (Y axis), problems that have a slightly wider field of activity, such as across sections or divisions, are selected. The autonomy of the activities (Z axis) increases and activities are autonomously implemented and managed in the team. The members, who have a sense of success at this second stage, begin to work more positively.

Step three: The contribution to performance stage

At the third stage of activities, connection with daily work is the key point. Referring to the objectives of the team's own division, problems are identified and solutions to problems are sought by participating in small-group activities.

At this stage, the business operation of the division and small-group activities are not discussed as separate subjects. It is necessary to coordinate closely with the management in implementing small-group activities. Since the themes selected are the results of breaking down the objectives of the division, carrying out small-group activities in itself cannot be a factor of selection. The scope of themes (Y axis) is extended greatly. For some themes, it is impossible for the problems to be solved by the members of the team. Therefore, people in charge of processes that occur before or after the process in question will join the group as members.

Step four: The participation in management stage

At the fourth stage, small-group activities do not exist independently; rather, the way in which daily tasks are performed takes on the very essence of small-group activities—independence and autonomy. In the current business situation, where the environment changes rapidly and business itself is diversifying, the performance of daily tasks requires not only exceptional performance of duties, but also an attitude of autonomously responding to changes at all times. The role of senior management is to formulate management policy so as to set an environment in which autonomous action will not be hampered.

As for the small-group activities in stage four, these activities (X axis) will extend to other divisions and affiliated companies when necessary, and the scope of themes (Y axis) is also extended to all the fields defined by the company policy.

Appendix

The 30th anniversary of the founding of ZD activitiy occurred in 1997. To commemorate the event, the Japan Management Association established a planning committee for the 30th anniversary convention in October of 1996. The committee prepared a proposal for the shape of ZD activities as we approach the 21st century. (Chapter 8 summarized the ideas in that proposal.) The proposal was the result of six months of discussions of the planning committee with a group of experts. The following three ideas highlight the main themes of the proposal, which looks at the course of small-group activities as we move toward the 21st century.

1. Recognition and definition of *diversity* of small-group activities
2. Reconfirmation of the *position* of small-group activities
3. Recognition and definition of the *stages of development* of small-group activities

This proposal was made possible thanks to the enthusiasm and effort of the members of the planning committee for the 30th anniversary convention of ZD activity. We would like to express our deepest appreciation to the committee members (please see the list of members on the following page) and hope that their proposal will contribute to the development of future small-group activities.

ZD 30th Anniversary Convention

(A national gathering to discuss diversified small-group activities)

Planning Committee

Name	Company*	Title and Division Name
Chairman		
Masaaki Naganuma	NEC Corp.	General Manager, Customer Satisfaction and Quality Management Promotion Division
Members		
Kazutake Sawada	Canon Inc.	Staff Manager, G-CPS Administration Dept., Production Management Administration Division
Koji Umeda	Kokusai Electric Co., Ltd.	Manager, "GO-Sine" Promotion Group
Takeshi Shibuya	Sanyo Electric Co., Ltd.	Manager, Tokyo Plant, Management Office
Hitoshi Teramoto	Nippon Steel Corp.	Senior Manager, Technical Administration and Planning Dept., Technical Administration and Planning Division
Kansyu Houki	Sumitomo 3M Ltd.	Manager, Corporate Quality Dept.
Atsuo Ito	Sumitomo Electric Industries Ltd.	General Manager, Plant and Production Systems Engineering Division
Mamoru Sumimoto	Sony Corp.	General Manager, Reliability and Quality Assurance Dept., Customer Satisfaction Center
Kazuo Sawada	Toshiba Corp.	Senior Specialist, Quality Assurance Center, Productivity Division
Tadao Furuo	Toshiba Engineering Corp.	Chief Specialist, Strategic Business Planning Division
Hiroaki Onishi	Toyo Communication Equipment Co., Ltd.	General Manager, Quality Assurance Division
Makoto Matsui	Nichirei Corp.	Deputy Manager, Human Resources Division
Yoshikane Tanaka	Japan Airlines	Senior Director, Engineering and Maintenance
Masao Saito	Nippon Signal Co., Ltd.	Senior Manager, Quality Control Division
Yasuo Saka	NEC Corp.	Chief Manager, Customer Satisfaction and Quality Management Promotion Division
Kenichi Hiramoto	East Japan Railway Co.	Deputy Manager, Personnel Dept.
Masao Arai	Hitachi Zosen Corp.	Manager, Corporate Planning Dept.
Hidehiro Yashiro	Hitachi Electronics Services Co., Ltd.	Center Manager, Management Innovation Promotion Center
Sadahiko Kamatani	Fuji Xerox Co., Ltd.	Staff Manager, New Xerox Movement Office
		(As of June 1997)

(* in Japanese *Kana* alphabet order)

About the Author

Shigehiro Nakamura graduated from Waseda University with a master's degree in metal engineering. He joined Hitachi Metal in 1970 and introduced and deployed companywide engineering during his first eleven years with the company. Mr. Nakamura also built a pilot JIT plant as well as a pilot FA line at another plant and improved the subcontractor systems. He went on to promote VEC and CIM throughtout the company.

More recently, Mr. Nakamura managed the building and startup of AAP St. Mary in the United States. He was also manager of process/equipment control and system improvement, and in charge if the CIM project. He joined the Japan Management Association in 1991 as a Total Productivity Management consultant and instructor for JMA's management school. Mr. Nakamura has worked extensively with many companies. In recent years. Mr. Nakamura has partnered with Productivity Europe to bring Japanese management techniques to Europe. It is through Productivity Europe that Mr. Nakamura's writings about the go-go tools were first translated into English and published.

Books from Productivity Press

Productivity Press publishes books that empower individuals and companies to achieve excellence in quality, productivity, and the creative involvement of all employees. Through steadfast efforts to support the vision and strategy of continuous improvement, Productivity Press delivers today's leading-edge tools and techniques gathered directly from industry leaders around the world. Call toll-free (800) 394-6868 for our free catalog.

20 Keys to Workplace Improvement (Revised Edition)

Iwao Kobayashi

The 20 Keys system does more than just bring together twenty of the world's top manufacturing improvement approaches—it integrates these individual methods into a closely interrelated system for revolutionizing every aspect of your manufacturing organization. This revised edition of Kobayashi's bestseller amplifies the synergistic power of raising the levels of all these critical areas simultaneously. The new edition presents upgraded criteria for the five-level scoring system in most of the 20 Keys, supporting your progress toward becoming not only best in your industry but best in the world.
ISBN 1-56327-109-5 / 302 pages / $50.00 / Order 20KREV-B8003

40 Top Tools for Manufacturers
A Guide for Implementing Powerful Improvement Activities

Walter Michalski

We know how important it is for you to have the right tool when you need it. And if you're a team leader or facilitator in a manufacturing environment, you've probably been searching a long time for a collection of implementation tools tailored specifically to your needs. Well, look no further. Based on the same principles and user-friendly design of the *Tool Navigator's The Master Guide for Teams*, here is a group of 40 dynamic tools to help you and your teams implement powerful manufacturing process improvement. Use this essential resource to select, sequence, and apply major TQM tools, methods, and processes.
ISBN 1-56327-197-4 / 160 pages / $25.00 / Order NAV2-B8003

Becoming Lean
Inside Stories of U.S. Manufacturers

Jeffrey Liker

Most other books on lean management focus on technical methods and offer a picture of what a lean system should look like. Some provide snapshots of before and after. This is the first book to provide technical descriptions of successful solutions and performance improvements. The first book to include powerful first-hand accounts of the complete process of change, its impact on the entire organization, and the rewards and benefits of becoming lean. At the heart of this book you will find the stories of American manufacturers who have successfully implemented lean methods. Authors offer personalized accounts of their organization's lean transformation, including struggles and successes, frustrations and surprises. Now you have a unique opportunity to go inside their implementation process to see what worked, what didn't, and why. Many of these executives and managers who led the charge to becoming lean in their organizations tell their stories here for the first time!
ISBN 1-56327-173-7 / 350 pages / $35.00 / Order LEAN-B8003

Building a Shared Vision

A Leader's Guide to Aligning the Organization

C. Patrick Lewis

This exciting new book presents a step-by-step method for developing your organizational vision. It teaches how to build and maintain a shared vision directed from the top down, but encompassing the views of all the members and stakeholders, and understanding the competitive environment of the organization. Like ***Corporate Diagnosis***, this books describes in detail one of the necessary first steps from *Implementing a Lean Management System*: visioning.
ISBN 1-56327-163-X / 150 pages / $45.00 / Order VISION- B8003

Caught in the Middle

A Leadership Guide for Partnership in the Workplace

Rick Maurer

Managers today are caught between old skills and new expectations. You're expected not only to improve quality and services, but also to get staff more involved. This stimulating book provides the inspiration and know-how to achieve these goals as it brings to light the rewards of establishing a real partnership with your staff. Includes self-assessment questionnaires.
ISBN 1-56327-158-3 / 258 pages / $30.00 / Order CAUGHT-B8003

Corporate Diagnosis

Setting the Global Standard for Excellence

Thomas L. Jackson with Constance E. Dyer

All too often, strategic planning neglects an essential first step and final step-diagnosis of the organization's current state. What's required is a systematic review of the critical factors in organizational learning and growth, factors that require monitoring, measurement, and management to ensure that your company competes successfully. This executive workbook provides a step-by-step method for diagnosing an organization's strategic health and measuring its overall competitiveness against world class standards. With checklists, charts, and detailed explanations, ***Corporate Diagnosis*** is a practical instruction manual. Detailed diagnostic questions in each area are provided as guidelines for developing your own self-assessment survey.
ISBN 1-56327-086-2 / 115 pages / $65.00 / Order CDIAG-B8003

Do it Right the Second Time

Benchmarking Best Practices in the Quality Change Process

Peter Merrill

Is your organization looking back on its quality process and saying "it failed"? Are you concerned that TQM is just another fad, only to be replaced by the next improvement movement? Don't jump ship just yet. Everyone experiences failures in their quality improvement process. Successful organizations are different because they learn from their failure: They do it right the second time. In this plain-speaking, easy-to-read book, Peter Merrill helps companies take what they learned from their first attempts at implementing a quality program, rethink the plan, and move forward. He takes you sequentially through the activities required to lead a lasting change from vision to final realization. Each brief chapter covers a specific topic in a framework which leads you directly to the issues that concern your organization.
ISBN 1-56327-175-3 / 225 pages / $27.00 / Order RSEC-B8003

Feedback Toolkit

16 Tools for Better Communication in the Workplace

Rick Maurer

In companies striving to reduce hierarchy and foster trust and responsible participation, good person-to-person feedback can be as important as sophisticated computer technology in enabling effective teamwork. Feedback is an important map of your situation, a way to tell whether you are "on or off track." Used well, feedback can motivate people to their highest level of performance. Despite its significance, this level of information sharing makes most managers uncomfortable. ***Feedback Toolkit*** addresses this natural hesitation with an easy-to-grasp 6-step framework and 16 practical and creative approaches for giving and receiving feedback with individuals and groups.
ISBN 1-56327-056-0 / 109 pages / $12.00 / Order FEED-B8003

Handbook for Personal Productivity

Henry E. Liebling

A little book with a lot of power that will help you become more successful and satisfied at work, as well as in your personal life. This pocket-sized handbook offers sections on personal productivity improvement, team achievement, quality customer service, improving your health, and how to get the most out of workshops and seminars. Special bulk discounts are available (call for more information).
ISBN 1-56327-131-1 / 128 pages / $5.00 paper / Order PERS-B8003

The Idea Book

Improvement Through TEI (Total Employee Involvement)

Japan Human Relations Association

At last, a book showing how to create Total Employee Involvement (TEI) and get hundreds of ideas from each employee every year to improve every aspect of your organization. Gathering improvement ideas from your entire workforce is a must for global competitiveness. ***The Idea Book***, heavily illustrated, is a hands-on teaching tool for workers and supervisors to refer to again and again. Perfect for study groups, too.
ISBN 0-915299-22-4 / 232 pages / $40.00 / Order IDEA-B8003

Implementing a Lean Management System

Thomas L. Jackson with Karen R. Jones

Does your company think and act ahead of technological change, ahead of the customer, and ahead of the competition? Thinking strategically requires a company to face these questions with a clear future image of itself. ***Implementing a Lean Management System*** lays out a comprehensive management system for aligning the firm's vision of the future with market realities. Based on hoshin management, the Japanese strategic planning method used by top managers for driving TQM throughout an organization, Lean Management is about deploying vision, strategy, and policy to all levels of daily activity. It is an eminently practical methodology emerging out of the implementation of continuous improvement methods and employee involvement. The key tools of this book build on multiskilling, the knowledge of the worker, and an understanding of the role of the new lean manufacturer.
ISBN 1-56327-085-4 / 182 pages / $65.00 / Order ILMS-B8003

Kaizen Teian 1

Developing Systems for Continuous Improvement Through Employee Suggestions

Japan Human Relations Association (ed.)

Especially relevant for middle and upper managers, this book focuses on the role of managers as catalysts in spurring employee ideas and facilitating their implementation. It explains how to run a proposal program on a day-to-day basis and outlines the policies that support a "bottom-up" system of innovation and defines the three main objectives of kaizen teian: to build participation, develop individual skills, and achieve higher profits.
ISBN 1-56327-186-9 / 217 pages / $30.00 / Order KT1P-B8003

Secrets of a Successful Employee Recognition System

Daniel C. Boyle

As the human resource manager of a failing manufacturing plant, Dan Boyle was desperate to find a way to motivate employees and break down the barrier between management and the union. He came up with a simple idea to say thank you to your employees for doing their job. In ***Secrets to a Successful Employee Recognition System***, Boyle outlines how to begin and run a 100 Club program. Filled with case studies and detailed guidelines, this book underscores the power behind thanking your employees for a job well done.
ISBN 1-56327-083-8 / 250 pages / $25.00 / Order SECRET-B8003

Tool Navigator

The Master Guide for Teams

Walter J. Michalski

Are you constantly searching for just the right tool to help your team efforts? Do you find yourself not sure which to use next? Here's the largest tool compendium of facilitation and problem solving tools you'll find. Each tool is presented in a two to three page spread which describes the tool, its use, how to implement it, and an example. Charts provide a matrix to help you choose the right tool for your needs. Plus, you can combine tools to help your team navigate through any problem solving or improvement process. Use these tools for all seasons: team building, idea generating, data collecting, analyzing/trending, evaluating/selecting, decision making, planning/presenting, and more!
ISBN 1-56327-178-8 / 550 pages / $150.00 / Order NAV1-B8003

TO ORDER: Write, phone, or fax Productivity Press, Dept. BK, P.O. Box 13390, Portland, OR 97213-0390, phone 1-800-394-6868, fax 1-800-394-6286.
Outside the U.S. phone (503) 235-0600; fax (503) 235-0909
Send check or charge to your credit card (American Express, Visa, MasterCard accepted).

U.S. ORDERS: Add $5 shipping for first book, $2 each additional for UPS surface delivery. Add $5 for each AV program containing 1 or 2 tapes; add $12 for each AV program containing 3 or more tapes. We offer attractive quantity discounts for bulk purchases of individual titles; call for more information.

ORDER BY E-MAIL: Order 24 hours a day from anywhere in the world.
Use either address:

To order: service@ppress.com

To view the online catalog and/or order: http://www.ppress.com/

QUANTITY DISCOUNTS: For information on quantity discounts, please contact our sales department.

INTERNATIONAL ORDERS: Write, phone, or fax for quote and indicate shipping method desired. For international callers, telephone number is 503-235-0600 and fax number is 503-235-0909. Prepayment in U.S. dollars must accompany your order (checks must be drawn on U.S. banks). When quote is returned with payment, your order will be shipped promptly by the method requested.

NOTE: Prices are in U.S. dollars and are subject to change without notice.